Copyright © 2021 by Jack T.R. Wilkin
Creative Commons (CC BY)
ISBN: 978-1-6671-1873-4

Contents

Introduction 3
History 4
Palaeogeography 6
Conditions in the Basins 11
Lithology and Depositional Models 15
Taphonomy 18
Microorganisms 23
Plants 25
Reef Builders 26
Soft-bodied animals 26
Arthropods 28
Bivalves 31
Cephalopods 33
Fish 39
Reptiles 41
Dinosaurs 49
Archaeopteryx 54
Final Thoughts 58
Bibliography 59

Introduction

The fossil record is notoriously incomplete, giving us only glimpses into the history of life on our planet throughout geological time. Konservat-Lagerstätten provides a rare opportunity to view ancient ecosystems in their entirety. Most advances in our understanding of prehistoric life come from exceptionally persevered biota- Lagerstätten. The term Lagerstätten is derived from the German mining term referring to a stem rich in ore, which Seilacher *et al.* (1985) compared to highly fossiliferous strata. Fossil Lagerstätte is best translated into English as "fossil bonanza", but as scientists, we must preserve the illusion of professionalism.

The Late Jurassic lithographic limestones, or Plattenkalks of southern Germany have yielded some of the most beautiful and scientifically important fossils in the world. The limestones, termed Plattenkalks which translates from German to *flat chalk*, are a textbook and widely studied example of a Konservate-Lagerstätten. The quality of the preservation means that details such as soft tissues and even colour patterns are known.

This book aims to provide a short introduction to the Solnhofen limestones, or Plattenkalks, of southern Germany. The Plattenkalks are concentrated in the south German states of Baden-Württemberg and Bavaria. The area being examined in the following book is the Southern Franconian Alb (Südliche Fränkische Alb), which lies roughly halfway between the Bavarian cities of Munich to the south and Nuremberg to the north (Figure 1).

Fig. 1. Locations of the various Plattenkalk sites in the Franconian and Swabian Alb. Redrawn from from Lomax *et al.* (2017).

History

The people of central Europe have used the Solnhofen Limestones for millennium. In caves dating to the Late Palaeolithic Magdalenian culture 17,000 to 12,000 years ago, Plattenkalk slabs were used as decoration. Early humans even scratched drawings on the limestone slabs and painted them with red ochre (Conard and Malina, 2010, 2011; Wolf et al., 2018).

From Roman times onwards, the limestones were used as roofing tiles and to line public baths due to their abundance, flat regular bedding and ease of extraction. The Romans also used the stone as a facing stone and tablets for inscriptions (Barthel et al., 1990).

During the Medieval period, the Plattenkalks were still being used in house construction, especially for use as floor and roof tiles. The stone was also a valuable export. For example, the Byzantine *Hagia Sophia* built between 532 and 537 AD on the orders of the Byzantine Emperor Justinian I, had a mosaic floor made from Plattenkalk tiles from Kelheim. Plattenkalk tablets are also common in Medieval churches across southern Germany (Barthel et al., 1990).

The main limestone quarries are concentrated in the western part of the Franconian Alb because the limestone here is purer and therefore more highly prized as a building stone and for use in lithography (Crook, 1894; Barthel et al., 1990). Major quarry sites are dotted across the region, especially around the village of Solnhofen and the Baroque town of Eichstätt each one having a fantastic fossil museum displaying local finds.

More recently they have been used for lithography, a printing process used for books and paintings. The story of the discovery of lithology goes that in 1793 Alois Senefelder wanted to make a list of items he wanted a washerwoman to collect but did not have any paper to hand. So, he decided to write on a limestone slab. To make the message clearer, he washed the tile with a weak acid. The acid was only absorbed and dissolved by the limestone not coated in ink, enhancing the letters. When a sheet of paper was later placed on the slab, it produced a perfect mirror image of the inked design. Senefelder immediately knew that he could use this method to produce art quickly and cheaply so wasted no time in perfecting the technique. Lithographic prints were incredibly popular and allowed art to be owned by ordinary people, not just the Bavarian elite. Newer methods in the Twentieth Century soon replaced lithography, but lithographic prints are still being produced to this day (Barthel *et al.*, 1990).

Because of the constant quarrying over the last few thousand years, it is not surprising that workers have been discovering fossils since the Stone Age. In the Neolithic, fossils were kept as ornaments or were perhaps used as amulets for religious purposes. Slabs with fossils, although highly prized now and Solnhofen fossils are one of the main reasons I became interested in palaeontology in the first place, make for lousy building material and fossils were often discarded. Quarry owners allowed workers to keep fossils they found and so quarrymen built extensive fossil collections over the years.

During the Scientific Revolution people became more interested in fossils and trade began with quarrymen and farmers selling fossils to collectors both in Bavaria (which was still an independent kingdom at the time) and aboard. Johann Jakob Baier published the first major scientific paper concerning Solnhofen material in *Oryktographia Norica* in 1708.

In 1861 an isolated feather was discovered near Solnhofen renewed interest in the Plattenkalks (Figure 2). Later that same year an almost complete skeleton was later named the London Specimen was discovered by quarry workers, lacking only the skull (Figure 3). The specimen displayed a mixture of avian and reptilian features and would go down in history as the first bird and the first discovered transitional fossil. The discovery of this animal, later named *Archaeopteryx lithographica* (von Meyer, 1861), could not have been come at a better time, found just two years after the publication of Charles Darwin's *The Origin of Species* (1859).

Archaeopteryx was heralded as Darwin's predicted "missing link". Thomas Huxley, one of Darwin's most vocal supporters, concluded that *Archaeopteryx* was a transitional form between dinosaurs and birds in a landmark paper published in 1868. Huxley's theories would later be expanded upon by later authors during the Dinosaur Renaissance (e.g., Ostrom, 1969, 1973, 1974; Bakker, 1986), discoveries in China (Qiang and Shu-an, 1996; Chen *et al.*, 1998; Xu and Norell, 2004; Feduccia, 2012), and studies on dinosaur metabolisms (Amiot *et al.*, 2016) and respiratory systems (Ruben *et al.*, 1997; O'Connor and Claessens, 2005; Sereno *et al.*, 2008) that now the majority of scientists agree that birds are the descendants of small theropod dinosaurs.

Fig. 2. Single fossil feather found 1861. WikiMedia cc.4.0

Fig. 3. Replica of the London Specimen of Archaeopteryx. WikiMedia cc.4.0

Palaeogeography

The Jurassic of Southern Germany, predominantly in Baden-Württemberg and Bavaria, formed part of the epicontinental shelf of the north-western Tethys Ocean (Schweigert, 1998b; Keupp *et* al., 2007; Pieńkowski *et al.*, 2008 – for a review of the geological history of the Tethys Ocean see Stow, 2010). Sedimentary features suggest that the Plattenkalks were formed within isolated basins surrounded by extensive bioherms. During the Late Jurassic, Europe was an archipelago surrounded by a shallow sea located over the tropics (Figure 4). These basins are considered to have been deep areas within a relatively shallow epicontinental sea. Depending on sea-level, the various basins were more-or-less connected and, thus, well connected to the sea

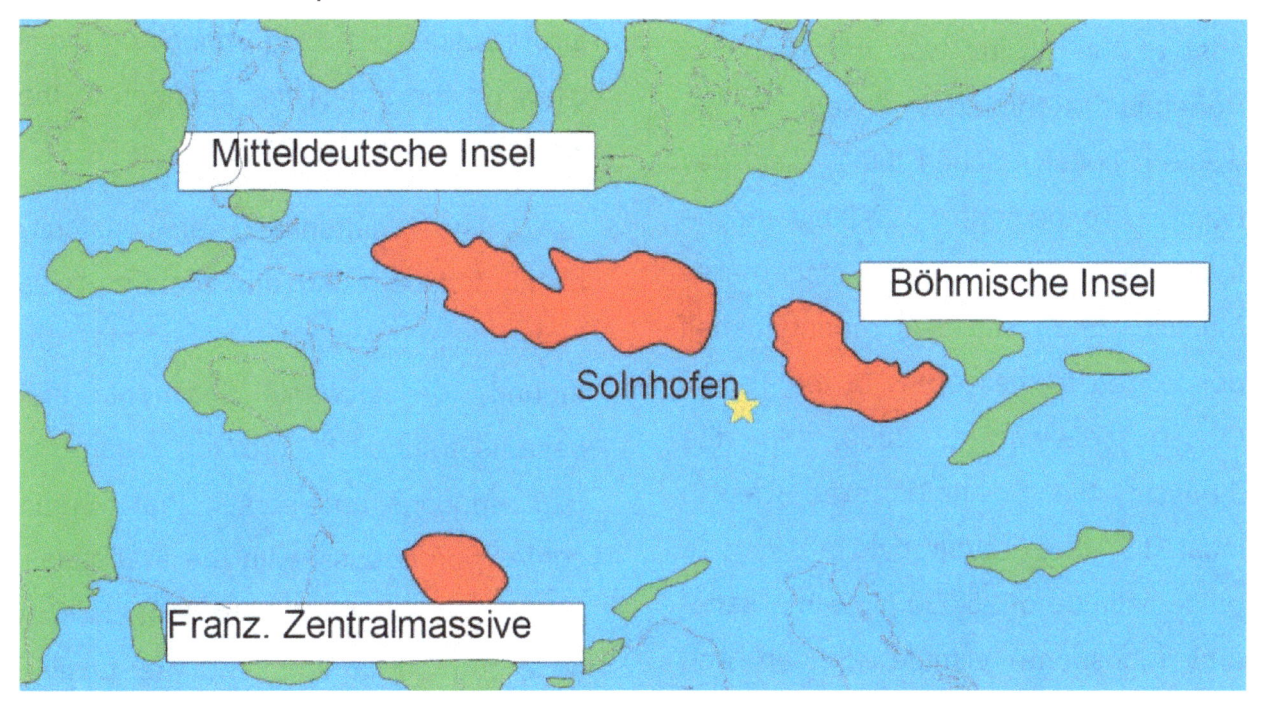

Fig. 4. A map of Central Europe during the Late Jurassic (c.150 million years ago). Blue= sea, light green= islands, red=islands with corresponding labels. Redrawn from Wilkin (2020).

Early depositional models concluded, incorrectly, that the Southern Franconian Alb was an extensive mudflat where sediments were brought in by storms and left to dry out in the sun. However, there is no evidence to support subaerial exposure due to the lack of diagnostic sedimentary features such as desiccation phenomena or evaporitic sequences. It was quite the contrary, with the regular and tabular nature of the Plattenkalk, suggesting deposition in a deep, quiet, protected environment under permanent subaqueous conditions (Barthel *et al.*, 1990). A present-day equivalent would be the Bahamas (Flügel, 2013). However, it is worth remembering when dealing with the

Plattenkalk that each basin varies in terms of palaeoenvironment. As such, each basin should be studied individually, and generalisations should be avoided (Ebert *et al.*, 2015). To the east of the basins was the Island of Bohemia (Böhmische Insel) whose western coast reached the present-day Danube. To the north was Mitteldeutsche Insel. In the south, near modern-day Munich, the shallow marine shelf descended to the depths of the Tethys Ocean (Wellnhofer, 2009). In the Jurassic, the Solnhofen area was in located the subtropics with a palaeolatitude of 25°-30°N, with a semi-arid monsoonal climate (Selden and Nudds, 2012). The Tropics were somewhat warmer in the Mesozoic than today (Bender, 2013). The Late Jurassic of Southern Germany was a part of the Franconian-South Bavarian Carbonate Platform, built by sponge-microbial bioherms (Viohl, 1996). A bioherm is a tabular reef rather than a dome-shaped one (Nichols, 2009) that grew on pre-existing submarine highs of Oxfordian age before becoming more expansive throughout the Early Kimmeridgian, and reaching their maximum extent in the late Middle Kimmeridgian and Tithonian. A marine regression during the Late Jurassic is indicated by ooid and peloid carbonate sands as well as by stromatolites replacing sponges. Corals grew on the top of the sponges in the Kelheim area (Viohl, 1996).

The Plattenkalk are patchily distributed, surrounded by more massive biohermal limestones. Bioherms are mounds of carbonate sediments that became trapped by growing organisms (e.g. sponges and corals) that formed contemporaneously with the Plattenkalk (Barthel *et al.*, 1990). The Plattenkalk's were formed in small marine basins, generally less than 10km in diameter, within the carbonate platform (Ebert *et al.*, 2015; Figure 5). These basins are considered to have been areas of deep bathymetry within a relatively shallow epicontinental carbonate sea. Depending on sea-level, the various basins were more-or-less connected, thus well connected to the sea (Reisdorf and Wittke, 2012).

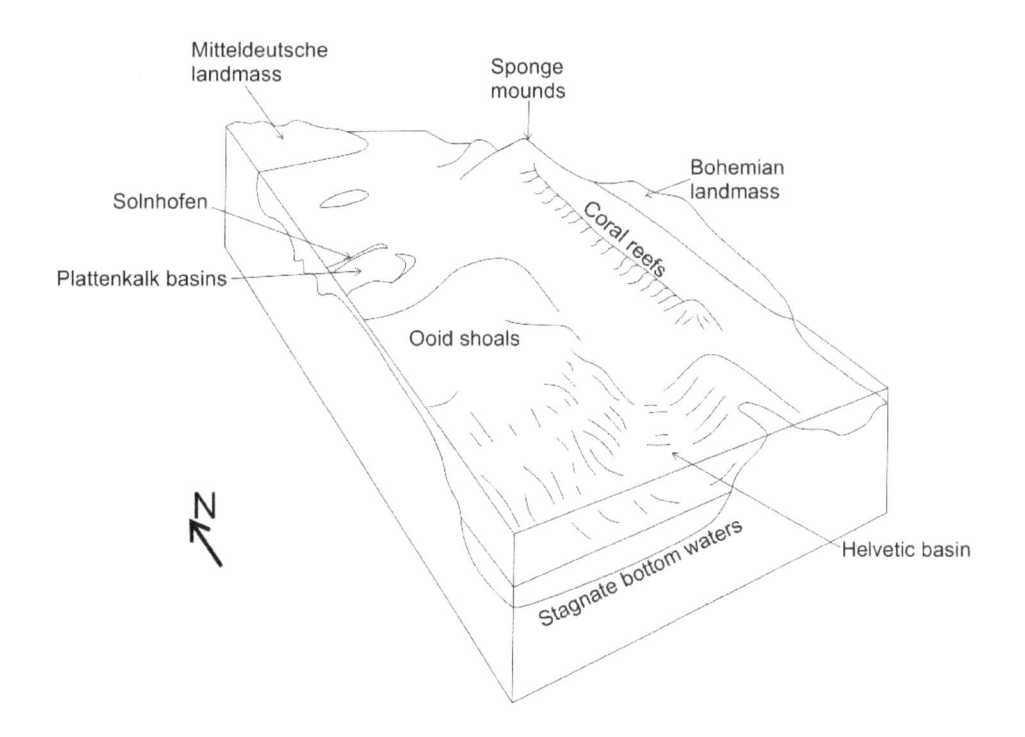

Fig. 5. Facies map of the Franconian Alb during the Tithonian. Redrawn from Wilkin (2018) originally from Barthel *et al.* (1990).

Marine transgressions had a significant effect on the palaeogeography of the Solnhofen region. Substantial portions of Central Europe were flooded during the Jurassic, including the area that would become Solnhofen. The coastline migrated towards the south and east. A resistant spar of land around the town of Landshut with a corresponding indentation known as the Regensberg embayment developed. Continued sea level rise and erosion meant the coastline breached, connecting the Tethys to the northern waters. Marine transgression in the Late Jurassic flooded Vindelicisch Land (Barthel *et al.*, 1990; Figure 6).

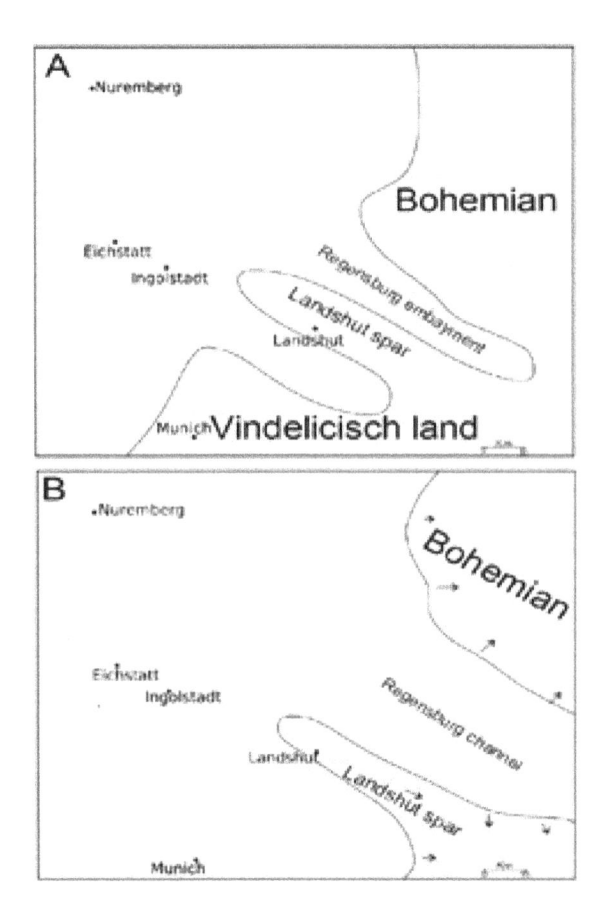

Fig. 6. Development of the coastline during the Jurassic. (A) Early Jurassic; (B) Late Jurassic. Redrawn from Wilkin (2018) originally from Barthel *et al.* (1990).

Fully marine deposition over the Southern Franconian Alb region did not start until the Jurassic. Dark shales and argillaceous limestones dominate the Early Jurassic, hence why the period is sometimes informally referred to as the Black Jurassic (Tanabe *et al.*, 1984; Jenkyns and Clayton, 1986), and sandstones dominate the Middle Jurassic. Marine transgressions during the Middle Jurassic and Late Jurassic connected the northern and southern waters of the Tethys Ocean (Barthel *et al*, 1990). There was a marine transgression during Oxfordian-Kimmeridgian but a regression in the Tithonian (Hallam, 2001).

The Plattenkalks became more widespread from the Tithonian. The sea level had dropped, and water exchange within the Tethys was restricted. Conditions in the Tethys were unfavourable for sponge growth and, consequently, corals began to recolonise the top of the sponge mounds forming a barrier to the east and patch reefs to the south and reefal debris beds extended (Barthel *et al.*, 1990; Schmid *et al.*, 2005).

It is now accepted that the Plattenkalks were formed underwater in a restricted lagoon within a vast shelf sea, protected by the Mitteldeutsche Insel to the north and by extensive coral reefs in the south and east, separating the basin from the Tethys Ocean. There were connections to the open sea as evidenced by the number of marine organisms similar to those found in modern tropical reefs. The surrounding waters were within the eutrophic zone, as shown by the substantial coral reefs. The basins were also near to the land as shown by the presence of terrestrials such as reptiles and insects.

Conditions in the Basins

The Solnhofen area lay within the subtropical, semi-arid zone. Evidence for this comes from sedimentological studies, isotopic measurements, and the palaeobotany. Limestones are the predominant lithology, suggesting a warm environment with an abundance of calcareous organisms. The rarity of terrestrial sediments indicates a lack of major rivers. (Barthel et al., 1990).

The Plattenkalks were formed in stable, low energy, lagoon environments. There is still some debate about the lagoon's salinity, however. The thin laminations of the Plattenkalk limestone have been interpreted as evidence for hypersaline, stagnant, and anoxic bottom waters, which would have been inhospitable to life (Selden and Nudds, 2012). The resulting lack of indigenous macro-organisms on the seafloor is one explanation for the lack of bioturbation in the Plattenkalks. The spectacular ichnofossils are known as "death-marches", or *mortichnia*, made by the horseshoe crab *Mesolimulus* (Figure 7) and the decapod crab *Mecochirus* (Figure 8), which often show the body fossil at the end, demonstrate the inhospitable nature of the lagoon bottom (Brenchley and Harper, 1998). Mortichnia, is a ecological class of ichnofossil first proposed by Seilacher (2007), are fascinating as they preserve both trace and trace maker together, which is incredibly rare. It isn't just *Mesolimulus* and *Mecocheirus* that leave such traces, others have been found as well, including prawns (Schweigert, 1998a) and fish (Schweigert et al., 2016).

(Page 12) Fig. 7. *Mesolimulus* death trail. Taken from Wilkin (2018).

(Page13) Fig. 8. *Mecochirus* death trail. Taken from Wilkin (2018).

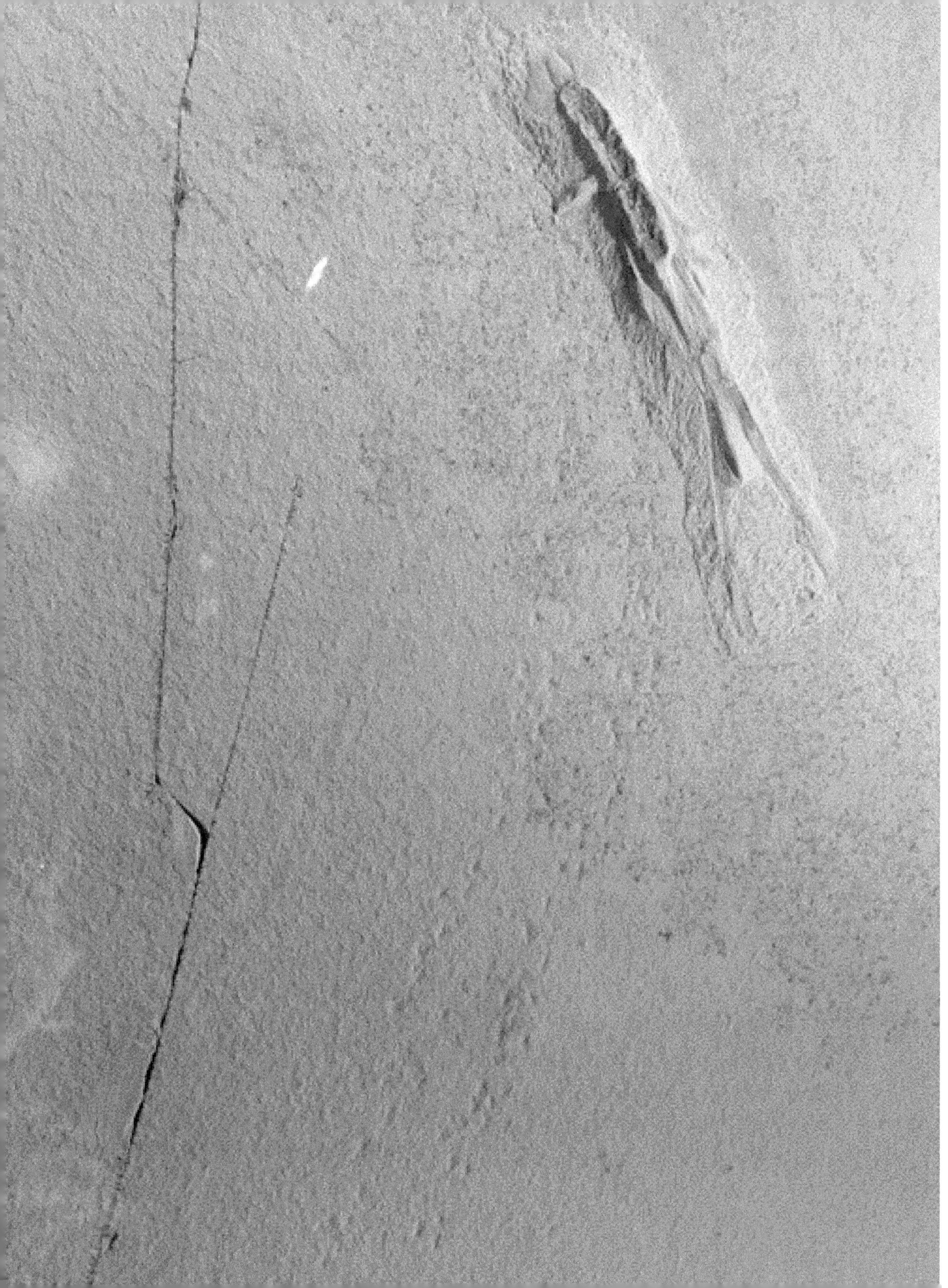

The land surrounding the Plattenkalk was hot and arid. Runoff was limited under such conditions, which explains the rarity of terrestrial biota and terrestrial sediments. Evaporation would have been substantial, but the influx of water during subtropical storms explains the lack of evaporites (Barthel *et al.*, 1990). As salinity increased brine pools delevoped at the bottom of the lagoon, making the water inhospitable to most life. Such conditions might explain the shrivelled appearance of fossil jellyfish, especially those from the Gungolding-Pfalzpaint area. High salt concentrations can also result in the pickling of organic remains because they slow microbial decay (Barthel *et al.*, 1990). The widely-accepted view is that the Plattenkalk formed beneath a halocline (a strong vertical salinity gradient) caused by high evaporation rates (Barthel *et al.*, 1990; Keupp *et al.*, 2007). Salinity in the basins is estimated to have ranged from 40-80% (Schwark *et al.*, 1998; Stevens *et al.*, 2014).

Any animals unfortunate enough to find themselves in this environment soon met their demise as clearly show by "death-marches". However, it should be noted that not all animals died because we fossils of animals that have shed their exoskeletons and then have walked away. Although deeply unpleasant for the poor inhabitants of ancient Bavaria, it is a godsend to palaeontologists as such conditions aid in the almost perfect preservation of these organisms. Even structures such as feathers and other soft tissues – including jellyfish - are preserved in exquisite detail.

At the same time, Rare Earth Element concentrations in organically-produced apatite from the Solnhofen Formation imply that bottom-water conditions were not anoxic (Kemp and Trueman, 2003). Also, the lack of stagnation deposits means that the basin was unlikely to have become anoxic before mixing and exchange of waters in the basins took place (Barthel *et al.*, 1990). The oxygen levels in the basins likely varied from normal marine to almost anoxic. Total anoxia was avoided due to the lack of organic production and frequent storms being in a fresh supply of water from the Tethys.

After subtropical storms introduced large amounts of seawater from the surrounding Tethys, the basins quickly returned to their near anoxic state for a couple of reasons. Firstly, the rate by which the oxygen being used by the decomposition of organics significantly outweighed the amount being produced by photosynthesisers living in the basins. Secondly, water with higher levels of salinity holds less oxygen in solution than those of lower salinity. The Plattenkalks have low levels of organic matter, due in part to the basins having very little productivity. Pelagic organisms may have lived in the better-aerated surface waters, at least for short periods (Selden and Nudds, 2012), which would explain the abundance of coprolites found in many Plattenkalks. There were times in which colonisation of the sediment was possible. For example, the lowermost part of the Solnhofen Formation was often bioturbated by worms and crustaceous. Both bioturbation and the coprolite evidence can be used to argue against total anoxia.

The plants from the surrounding islands include pteridosperms, Bennettitales, Lycophytes, cycads, sphenophytes, ferns, ginkgophytes, and conifers (Barbacka *et al.*, 2014). These plants only make up shrubs, and there is no evidence for large trees (Selden and Nudds, 2012) due to the lack of logs. Two theories have been proposed for this 1). The surrounding islands were devoid of large trees; 2) driftwood washed ashore and was merely not transported into the shallow waters of the basin (Wellnofer, 2009b).

Lithology and Depositional Models

The Plattenkalks progressively age towards the south and west, and the oldest deposits are described from Wattendorf in the Northern Franconian region. The Plattenkalk formations of Southern Franconia are, from oldest to youngest, the Dietfurt Formation, Arzberg Formation, Treuchtlingen Formation, Torleite Formation, Rögling Formation, Solnhofen Formation, the Mörnsheim Formation, Usseltel Formation, Rennertshofen Formation and the Neuburg Formation (Wilkin, 2018).

The Plattenkalks are comprised of a series of fine-grained laminated lithographic limestones. The beds are tabular and can be traced over long distances, several tens of metres or more, without any noticeable changes in appearance or lithology. The lithology is comprised of very pure, fine-grained (< 5 μm) micritic limestones with limited detrital content that form remarkably regular and laterally continuous beds. The limestone facies occur in two main types: flinz and fäule—the grain size peaks at 1-3 μm. The origins of these grains are unknown mainly due to recrystallised (Barthel *et al.*, 1990).

The flinz is finely grained and almost pure carbonate and seems to be severely altered by diagenetic processes. The material is likely the broken remains of coccoliths, but even when viewed under a Scanning Electron Microscope; it is difficult to ascertain any original forms. The fäule is more clay-rich and may have had an additional carbonate source produced *in-situ* by cyanobacterial mats. Most of the limestone layers are yellowish-grey in colour. However, distinctive reddish bands can be seen in some layers, produced by the oxidation of pyrite. The origins of the fäule are well established as well-preserved carbonate microfossils are often visible. The change in preservation between the two facies is accredited to the high clay contents of the fäule. The fäule beds weather faster than the flinz beds, making them recessive in profile. The way the facies split is also different; the flinz beds split more regularly along the bedding plane, and the fäule beds split into millimetre thin leaves. The mineral content of the fäule and flinz are presented in Figure 9.

Mineral	Flinz Beds	Fäule
$CaCO_3$	95-98%	77-87%
Clay minerals	Up to 3%	10-20%
Quartz	Up to 0.4%	3%

Fig. 9. Comparison of the mineral content of the flinz and fäule beds. Percentages are taken from Kemp and Trueman (2003).

As the carbonate oozes were compressed, the grains became welded together into a sturdy, supporting framework. Small unstable fragments of calcite and aragonite either recrystallised or were dissolved. This produced a $CaCO_3$ rich fluid that could precipitate as overgrowths or migrate and infill cavities with a blocky cement. Such blocky cement is non-fossiliferous (Barthel *et al.*, 1990).

Dendrites are inorganic ultra-fine-grained crystalline precipitates that often form branched shapes, superficially resembling fossil plants (Figure 10). Dendrites form along joint and bedding planes in the rock. They are deposited in pores in the rock and are cemented in place by Mn or Fe minerals (van Straaten, 1978). Dendrites are relatively common in the Plattenkalks.

Fig. 10. Manganese dendrites from the Solnhofen Formation. Scale in mm. Image curtsey of Mark A. Wilson (The College of Wooter).

Keupp (1977) proposed that the Flinz Beds formed under stagnated and saline conditions. Benthic cyanobacteria grew on the lagoon floor, forming a mat, which aided in the exceptional preservation of the fossils. Occasionally, storms would have caused an influx of seawater into the restricted lagoons. These storms would also have brought in organisms, including coccoliths which would have, after dying and sinking to the basin floor, contributed to the white coccolith laminae of the Flinz. The clay-rich beds formed when the water column was mixed, killing off the cyanobacteria and temporarily stopping carbonate production (Figure 11).

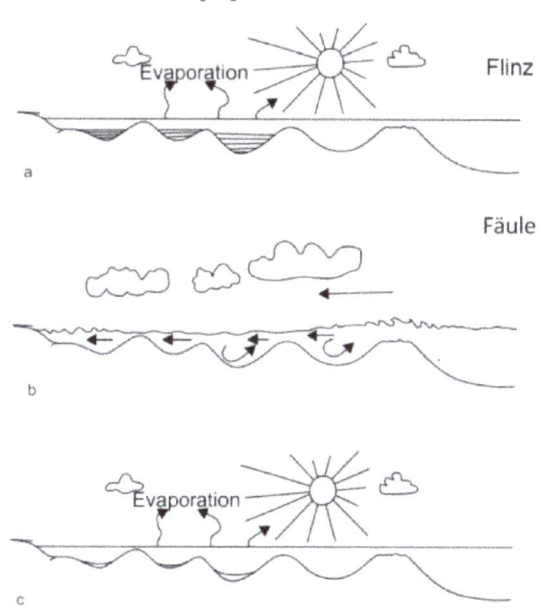

Fig. 11. Depositional model proposed by Keupp (1977). Redrawn from Wilkin (2018).

However, Keupp's theory has its contractors. Keupp's argument that stromatolites created the Flinz is questioned by Barthel *et al.* (1990). Stromatolites often cause irregular and wrinkled bedding planes but, as has been discussed previously, Plattenkalks have unusually uniform bedding surfaces. Also, the absence of pyrite and the low amounts of preserved organic matter both suggest that the Plattenkalk formed under conditions with low levels of organic productivity. In contrast, environments with elevated levels of biological productivity the sediments tend to become sulphidic, and organics are preserved more readily.

Instead Barthel *et al.* (1990) theorized that the carbonates in Plattenkalks came from the coral reefs that surrounded the basins. Periodically, perhaps related to seasonal variations in the climate, storms swept up the carbonate oozes from the seafloor and washed them over the reefs, flooding the lagoons (Figure 12). As such, the basin was not a permanent living space for the animals since it was likely a hostile environment in which organisms died shortly after being introduced (Petit and Khalloufi, 2012).

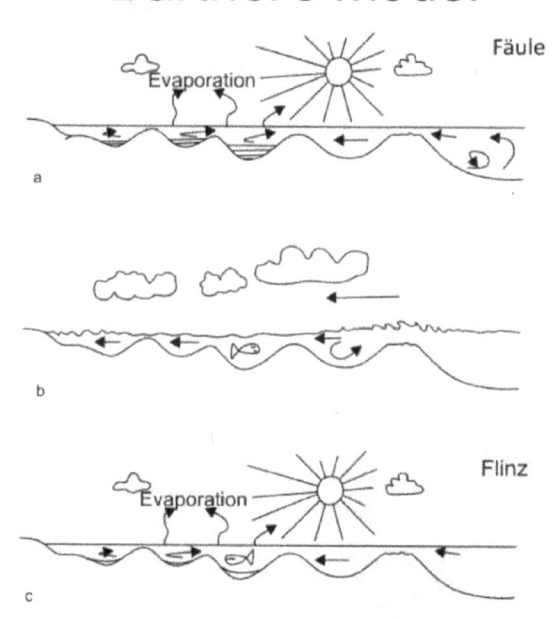

Fig. 12. Depositional model proposed by Barthel (1990). Redrawn from Wilkin (2018).

Taphonomy

The Plattenkalk are a Konservat Lagerstätten. Such deposits are notable for their extraordinary preservation as opposed to Konzentrat Lagerstätten which have abundant but poorly preserved fossils. The Plattenkalk arose from both stagnation and obrution. Stagnation deposits emerge from stagnated bottom waters and are commonly anoxic, making it inhospitable to benthic organisms. Most faunas present in these deposits are pelagic swimmers, but benthic organisms can sometimes be present when the bottom waters become dysaerobic or oxic. The other type of lagerstätte at work in the Plattenkalk are obrution deposits. Obrution deposits are formed by short-lived events such as storms that lead to rapid burial of organisms. In the Northern Tethys, this is caused by the subtropical storms that brought in flora and fauna from surrounding reefs and islands (Brenchley and Harper, 1998).

There are two taphonomic biases at work in the Plattenkalk basins; the first being transportation into the lagoon, and the second being the differential decay and dissolution of the sediment (Barthel *et al.*, 1990). Different authors give taphonomy different definitions and meaning (see Wilson, 1989 for an in-depth discussion) so it is worth specifying what this chapter will focus on. We will be considering necrolysis, the death and decomposition of the organism, and biostratinomy, the sedimentary history of the fossil before burial and diagenesis (Wilson, 1989; Tintori, 1992).

The taphonomy of fish and other vertebrate is extremely useful because they are made of many elements, making them sensitive to different depositional conditions (Tintori, 1992; Wilkin, 2019a). Schäfer (1972, p. 49) wrote: *"fish carcasses are more vulnerable to decay than other vertebrates. This has one advantage: once we are familiar with the restricted conditions that permit total preservation, our conclusions about the mode of death, living conditions, and fossil preservation of extinct species will be more precise."*

In marine and basin settings, vertebrates may be preserved as entire skeletons or, more commonly, as isolated skeletal fragments. The former is favoured when the carcass sinks below both the thermocline and halocline into a still anoxic environment (Benchley and Harper, 1998). As discussed previously, the Plattenkalks fit this model, and many complete skeletons have been found including those of fishes, reptiles (Witton, 2013), marine reptiles (Bardet and Fernandez, 2000), and birds (Kemp and Unwin, 1997). After death, the carcass may undergo disaggregation and dispersal (Brenchly and Harper, 1998).

In a study on the marine vertebrates of the Oxford Clay of Late Jurassic Britain, Martill (1985) identified five preservation styles. Although the palaeoenvironment of the Oxford Clay is vastly different from that of the Plattenkalks, the same principles are valid for both localities:

(1) Articulated skeletons: The entire skeleton is present and left in the same position as of the time of death, showing the genuine bone-to-bone relationships. In some cases, coprolites and even bones of prey and the found in the stomach cavities of fishes and other vertebrates (Korgan and Licht, 2013). Some fishes in the basins are even preserved in the act of eating prey too big for it which caused the fish to choke. If a specimen is articulated, it shows that the animal died at an unusually calm period in the basins as not to be disturbed by currents. Because a variety of fishes can be preserved as fully articulated skeletons, this condition is not caused by the fish's anatomy but rather the depositional setting.

(2) Disarticulated skeletons: The degree of articulation can vary greatly, even within the same horizon. For example, a skeleton may be completely disarticulated or remain mainly intact. In some cases, more unified elements, such as the vertebral column, may remain articulated while the rest of the skeleton is scattered. In many of the Plattenkalks, disarticulation was caused by currents and not scavengers. Evidence for current activity includes convex-up bivalve valves and palaeocurrent indicators such as fish scales in alignment (Wilkin, 2019a).

(3) Isolated elements: Isolated skeletal fragments, for example, teeth and scales probably dropped from decomposing carcasses within the water column. Other reasons for this may include scavengers, though this explanation is unlikely as the Plattenkalks could not support benthic organisms or disturbance from water currents.

(4) Worn elements: Weathering structures on bones indicate a prolonged period within the Taphonomic Active Zone and

exposure to the elements.

(5) Coprolites: Coprolites may provide insights into diets, but often they cannot be associated with an organism beyond generalisations.

Despite the rich diversity of fauna, the Plattenkalks do not represent a complete ecosystem. Instead, the biota is allochthonous; meaning they are derived from a range of habitats (Barthel *et al.*, 1990). When examining fossil assemblages, multiple issues need to be considered (Shipman, 1981): (1) life environments do not always equate to death environments, (2) abundance of fossils does not reflect abundance in original communities, (3) species found in same assemblages do not reflect sympatry of those species in life, and (4) absence of group does not indicate a lack in the original community.

The organisms found within the Plattenkalk inhabited the reefs that surrounded the basins. Organisms, both dead and alive, were likely brought in by monsoonal storms. The most common fossils are those of weak swimmers such as ammonites and small fishes, with stronger swimmers, such as vampyromorphs (vampire squids), and benthonic organisms (bivalves, brachiopods, and crustaceans), being proportionately rarer. Such a scenario is a clear example of a taphonomic bias. As a result, the Plattenkalks are unrepresentative of the broader ecology. Most of the biota was likely swept post-mortem, but occasionally living organisms may have been swept in as evidenced by rare ichnofossils (Barthel *et al.*, 1990). This type of assemblage is termed a taphocoenosis (Brenchley and Harper, 1998).

Some organisms may have floated to the surface waters due to an increase in buoyancy after death. In vertebrates, this is caused by bloating because gas builds up within the gut region during putrefaction. If disintegration occurs while the carcass is floating within the water column, various body parts can become dispersed. If the body loses buoyancy while entire, the body will be deposited as a complete skeleton (Brenchley and Harper, 1998). The air-filled chambers of cephalopods would have caused buoyancy until the chamber flooded with water. One fantastic example of this is an 8.5-meter-long drag mark complete with the preserved ammonite shell (belong to the species *Subplanites rueppellianus*) at its end (Lomax *et al.*, 2017). This trace is represented to two parallel ridges and furrows produced by the ribs of the ammonite shell as it drifted just above the sediment surface and did not reflect the behaviour of the living animal but rather the air-filled chambers that led to a dead, rotting ammonite that left behind an enormous mark – the longest of its type in the world.

Many of the fossil specimens show opisthotonic-posture. This occurs when the body arches over, giving the fossil a characteristic 'death-pose'. Several theories have been proposed for why this happens. (1) Perimortem death throes resulting from the affliction of the central nervous system resulting from brain damage and asphyxiation (Faux and Padian, 2007). Such a scenario might have arisen in the plattenkalk lagoon if the waters were anoxic, which is certainly seems to be the case at least at times (Barthel *et al.*, 1990). (2) elastic pull of ligaments as a result of muscles relaxing after death; (3) rigour mortise combined with the contraction of various muscles; (4) contraction of ligaments; (5) head and tail dangling while the carcass was floating in the water column; (6) anchoring of the skull in the sediment and the alteration of the post-cranial skeleton by water currents; (7) osmatic desiccation in a hypersaline environment (Faux and Padian, 2007). Irregular curvature indicates that the decaying soft tissues were weaker than during life and did not keep to the shape of the vertebral column (Bieńkowska-Wasiluk, 2010).

Soft tissues in the Plattenkalks are preserved in the following two ways (Frey and Martill 1998): (1) phosphatic replacements of soft tissues (permineralisation), which would be responsible for preserving muscle and other tissues in, for example, pterosaurs (Frey *et al.* 2003) and fishes (Wilby *et al.* 1995); (2) external moulds of soft tissues. This method is the more common and is the responsible for the preservation the feathers of Archaeopteryx and, because of the importance of this taxon, this mode of preservation has been investigated in some detail (e.g. Charig *et al.* 1986; Elzanowski 2002). The size of the sediment determines the quality of the external mould. The Plattenkalk is fine-grained (Barthel *et al.*, 1990), and, therefore, the amount of detail preserved is high. For example, the tail feathers of *Archaeopteryx* reveal details such as barbs and barbules.

Microorganisms

Small hollow spheres found within the fäule, and to a lesser extent the flinz, are thought to represent the remains of cyanobacteria. Cyanobacteria are very tolerant of harsh conditions being able to survive and even flourish to hypersaline waters. They are thought to have covered the bottom of the basins forming a microbial mat. This thin coating allowed delicate traces made by organisms and sediment to become preserved (Figure 13).

When the mat hardened it formed a thin calcium carbonate crust created as a by-product of respiration. The calcified cyanobacterial mats helped create the perfect conditions for exceptional preservation. In modern environments algal mats can cover a carcass within a few hours, allowing very little time for decay to set in. The cyanobacteria formed an envelope around the body, creating a reducing environment which slows down decomposition. When the soft tissues finally decayed, the sediment is already reasonably well consolidated resulting in the characteristic "pedestal preservation" seen in many soft-bodied fossils. It is worth reiterating that each basin represented a unique palaeoenvironment and cyanobacterial mats were not present in every Plattenkalk (Fürsich *et al.*, 2007).

Fig. 13. Feather-like structures in the sediment caused by grains being transported by currents over a cyanobacterial mat. BBM. Authors own work.

Both the fäule and flinz contain microfossils when examined under the Scanning Electron Microscope, though in the flinz they are often beyond recognition due to diagenesis. The first microfossils to be studied from the fäule were foraminifera by Groiss (1967). Forams are single-celled protists that first evolved near the end of the Precambrian with over 50,000 species currently known. Forams produce a test (or shell) commonly made out of $CaCO_3$, or they can be agglutinated, meaning that that use sediment particles to protect themselves. Forams are frequently used to reconstruct Cenozoic climates and as environmental indicators (for example, deep-sea benthic forams were instrumental in developing the "Zachos Curve"; Zachos *et al.*, 2008).

When the cyanobacterial mats decayed, they released nutrients that sustained a resisted population of ostracods and foraminifera. Foraminifera from the basins are rare and have very little diversity being represented by smooth, unornamented forms. (Figure 14). Low diversity of foraminifera is indicative of hostile conditions that only very few species could tolerate. A study by Robin *et al.* (2013) found evidence of nubeculariid forams encrusting a small lobster from a site near Eichstätt. This association is thought to have started when the host when it was still alive on the reefs surrounding the basins. The association represents the first and only known example of nubeculariids attaching to a mobile host.

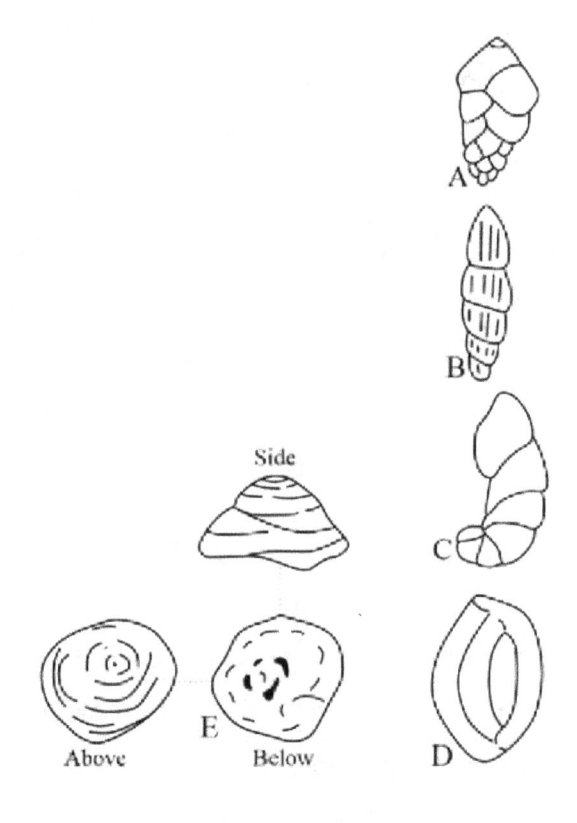

Fig. 14. Foraminifera from the Solnhofen Formation. (A) *Gaudryina bukowiensis*, 450µm; (B) *Nodosaria euglypha*, 1100µm; (C) *Marginulina distorta*, 550µm; (D) *Quinqueloculina egmontensis*, 490µm; (E) *Patellina feifeli*, 400µm. Redrawn from Barthel *et al.* (1990).

Plants

Plants recovered from the basins suggest the flora communities were not diverse. Both terrestrial and aquatic plants are known from the Plattenkalks. Most of the plant fragments are small, between 1-5 cm in length, and poorly preserved, mostly impressions but some retain their original carbon structure.

Aquatic plants include non-vascular phaeophytes (brown algae), were probably come along the Jurassic coastline. These algae are simple multicellular, macroalgae that could grow into giant kelp forests anchored to the seafloor.

The plants from the surrounding islands included pteridosperms, Bennettitales, Lycophytes, cycads, sphenophytes, ferns, ginkgophytes, and conifers; all are gymnosperms (that is-seed-bearing vascular plants; Figure 15). Low-lying conifers seem to be the most common plants. The arid conditions of the surrounding island were unsuitable for ferns, pteridophytes because ferns need damp habitats for reproduction.

There is no evidence for large trees due to the lack of logs. Two theories have been proposed for this 1). The surrounding islands were devoid of large trees; 2) driftwood washed ashore and was not transported into the shallow waters of the basin (Barthel *et al.*, 1990; Wellnofer, 2009b).

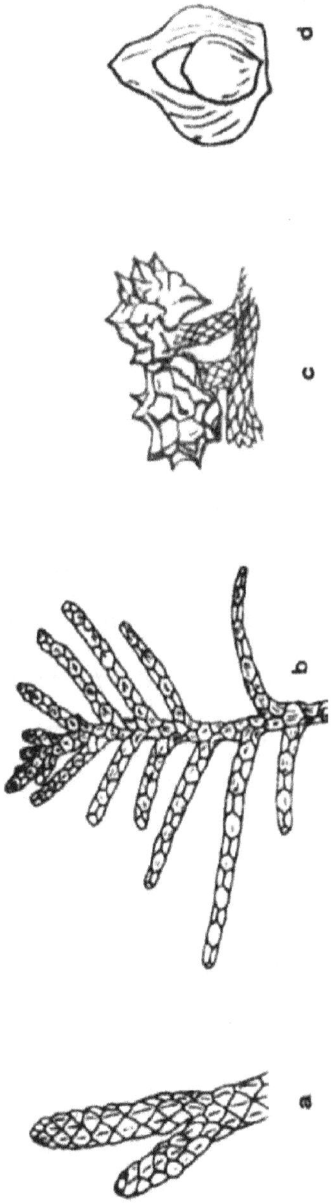

Fig. 15. Gymnosperms from the Plattenkalk. (A) *Brachyphyllum* twig. (B) *Palaeocyparis* twig. (C) *Arthrotaxites* cone. (C) Close-up of an araucarian cone scale. Redrawn from Barthel *et al.* (1990).

Reef Builders

Sponges are among the simplest metazoans. Sponge reefs helped trap sediments and built up the bioherms that separated the basins. Bioherms are mounds of carbonate sediments that became trapped by growing organisms (e.g. sponges and corals) that formed contemporaneously with the Plattenkalks. Spicules, structures that support the body, are used to identify different species (Barthel *et al.*, 1990).

Corals grew on top of the sponges forming extensive Jurassic reefs in Central Europe. Corals are colonial organisms. The Plattenkalks became more widespread during the Tithonian. The sea level had dropped, and water exchange within the Tethys was restricted. Conditions in the Tethys were unfavourable for sponge growth and, consequently, corals began to recolonise the top of the sponge mounds forming a barrier to the east and patch reefs to the south and reef debris beds extended (Wilkin, 2018).

Coral polyps form a symbiotic relationship with photosynthetic algae. There are examples of soft tissue preservation, from the Solnhofen Formation, which is incredibly rare (Bathel *et al.*, 1990).

Soft-bodied animals

Jellyfish belong to a group called cnidaria - which also includes corals. oldest jellyfish fossils appeared between 635 and 577 million years ago (Van Iten *et al.*, 2014) though they likely evolved much earlier, making them the oldest known form of multicellular life.

Jellyfish are extremely rare as fossils; their soft bodies, which are composed mostly of water, can only be preserved under very unusual conditions (see Schäfer, 1941; Hertweck, 1966; Müller, 1984, 1985; Barthel *et al.*, 1990; Bruton, 1991; Gaillard *et al.*, 2006). In Mesozoic deposits, they are best known from the Solnhofen Limestones (Kieslinger, 1939; Kuhn, 1938, 1961; Barthel *et al.*, 1990) and have also been found in similar deposits in Cerin, France (Gaillard *et al.*, 2006). The Solnhofen Plattenkalk is renowned for its jellyfish (scyphozoans), most of which come from the Gungolding-Pfalzpaint area. The most common jellyfish from the Plattenkalk is *Rhizostomites* (Figure 16).

It was a large jellyfish with a diameter greater than 50 centimetres (Barthel et al., 1990).

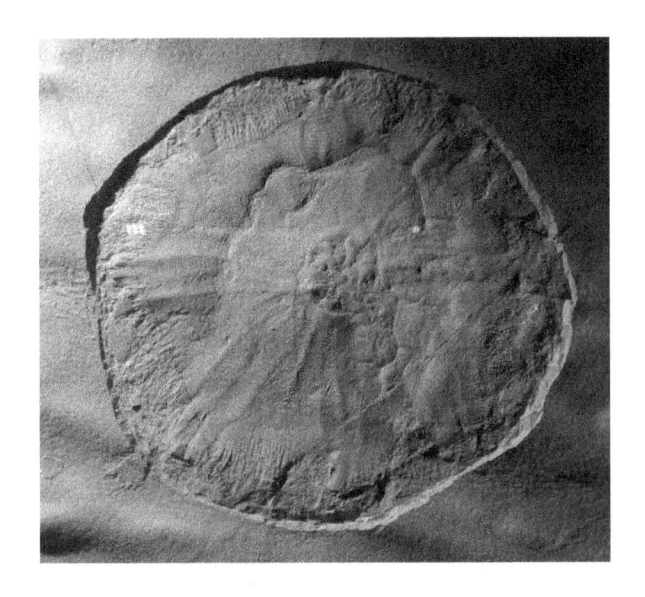

Fig. 16. Jellyfish, *Rhizostomites,* from the Gungolding-Pfalzpaint area. From Wilkin (2018).

The high salinity in the basins is responsible for the preservation of jellyfish due to the water being withdrawn from tissues osmotically. Such conditions would explain the shrivelled appearance of the fossil jellyfish. The preservation of some of the jellyfish is unusual because they contain sediment inside the bell, as shown by sections through jellyfish embedded in the sediment. The sediments must have been pumped in by the jellyfish themselves, presumably during their death throes (Barthel et al., 1990).

Another group of entirely soft-bodied animals from the Plattenkalks are annelids. Worms are not exclusively terrestrial; there are also marine species. The essential characteristics of the group are segmented bodies and bilateral symmetry. Annelids are soft-bodied which make them incredibly rare in the fossil record (see Parry, 2014 for a review of the annelid fossil record). However, some marine forms build themselves tube-shaped casing comprised of calcium carbonate, which is quite common as fossils. These tube-building worms, mostly given the generic name *serpula*, are still alive today and are encrusts in reefs during the Jurassic and still are today. Jurassic serpulids lived on driftwood and encrusted belemnite (Reolid et al., 2014) and ammonite shells (Seilacher, 1982). There are also rare examples of free-living marine annelids which lived either on or in the sediment (Barthel et al., 1990).

Arthropods

Arthropoda is the most species abundant phylum in the kingdom Animalia containing about 84% of all known animal species. Arthropods are invertebrates with an exoskeleton made from chitin, a segmented body and paired jointed appendages. Arthropoda includes insects, crustaceans, Chelicerata (which contains arachnids and horseshoe crabs and kin) and the now-extinct trilobites and eurypterids, which both went extinct at the end of the Permian (Clarkson, 1993).

The insect community of the Plattenkalks is diverse, with around two-fifths of the modern orders represented. Insects are terrestrial arthropods and are the most diverse group of living animals and account for the highest proportion of species from the Plattenkalks. The most notable being dragonflies (Odonata) and water boatman (Phasmida). Identification of these insects do genus, and species level is often difficult due to preservation, most specimens are impressions, lacking sufficient detail.

Dragonflies are characterised by their large eyes, short thorax, long abdomen and a pair of membranous wings. So perfect is the preservation in the basins that even the delicate vine venation on the wings is still present – even after 150 million years (Figure 17).

The long bodies and splayed-out legs of water boatman (Corixidae) enable them to skim across the surface of the clam basin waters effortlessly. *Chresmoda* may either have lived in the streams and ponds surrounding the basins or, as some modern forms, may have been marine.

Other insects include mayflies (Ephemeroptera), cockroaches (Blattoidea), locusts and crickets (Ensifera) Lacewings (Neuropterans) such as *Palparites deichmuelleri* from the Solnhofen Plattenkalk at Eichstätt, beetles (Coleoptera), and wasps (Hymenoptera).

Fig. 17. *Stenophlebia latreillei*. Image courtesy of Dr. Günter Bechly.

By far, the most common crustaceans from the Plattenkalks are the Decapoda - that is crabs (Figure 18), lobsters and prawns - is represented by 27 genera. As their name suggests, decapods have ten pairs of appendages. The final five pairs on the front half of the body termed the cephalothorax-so called because the head (cephallum) and thorax are fused- are used for locomotion. In many decapods, the frontmost appendages are used as claws called chelae which are used to manipulate food and help with feeding.

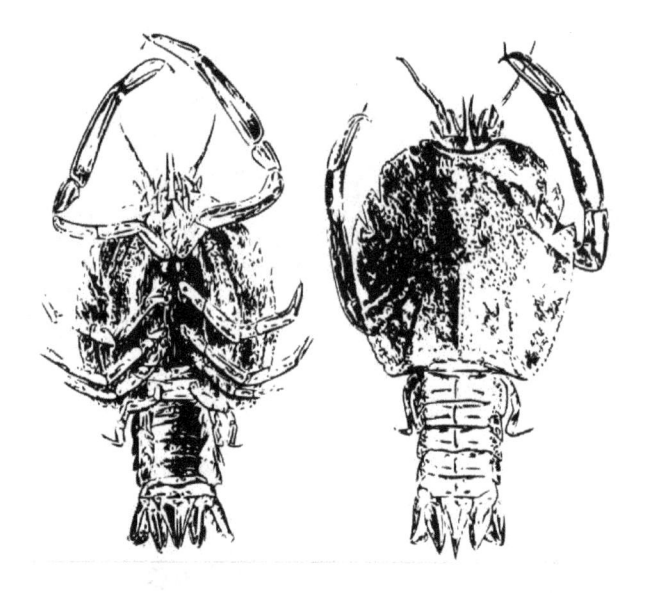

Fig. 18. *Cycleryon propinquus*. Redrawn from von Zittel and Eastman (1913).

The final arthropod phylum known from the south German Plattenklak is Chelicerata. Chelicerata is a diverse group which includes the arachnids (spiders, mites and scorpions) as well as horseshoe crabs. Despite being morphologically and ecologically diverse, they were paraphyletic being united by the following characteristics (Clarkson, 1993):

(1) An anterior prosoma made up of six segments which is equivalent to the fused head and thorax of the arthropods.

(2) The abdomen (opisthosoma) with 12 or fewer segments.

(3) A pair of jointed pinchers called chelicerae.

The only group of chelicerates for the Plattenkalks are horseshoe crabs (Subclass Xiphosura). Horseshoe crab lives in marine to brackish waters around the coastal shallows. During the spring high tides, they come ashore to lay their eggs (Rudloe, 1980) and it is presumed that extinct members did much the same. Horseshoe crabs are an ancient group, with the oldest fossils being over 445 million years old (Rudkin et al., 2008). Over the last nearly half a billion years, horseshoe crabs have changed very little, which makes them a prime example of a living-fossil (Figure 19). The genus known from the South German Jurassic is *Mesolimulus* (Barthel *et al.*, 1990;). As discussed previously, body fossils of horseshoe crabs are sometimes found at the end of trackways in so-called death trails which is used to demonstrate the inhibitability of the basins. The longest of these death trails was 9.7 meters (Lomax and Racay, 2012).

Fig. 19. *Mesolimulus walchi*. Image courtesy of Raimond Spekking (WikiMedia).

Bivalves

The Class Bivalvia is distinguishable from other molluscs by their laterally compressed bodies enclosed between two calcareous valves. The valves are connected on the dorsal side by an elastic horny ligament. In most cases, the animal is bilaterally symmetrical along the hinge margin: thus, the two valves are near perfect mirror images of each other. The function of the shell is to protect the soft edible part of the body (called the mantle) from predators. Modern examples of bivalves include cockles, mussels and oysters.

Bivalves are rare in the Plattenkalk but are more common than brachiopods. Cementing or byssal attached forms and the most commonly represented. Although soft tissues of bivalves seldom preserve, though are known in some exceptional specimens, their calcareous valves have a high preservation potential (Barthel *et al.*, 1990). Most of the examples are found as isolated valves (Wilkin, 2018) Figure 20). The reason for this is when bivalves die, the ligaments holding the two valves in place relax and the valves open to become separated and became scattered by currents (Gilinsky and Bennington, 1994).

Bed	Convex-up	Convex-down	Both valves
Ammoniten-Lagen	12	0	3
Ereste Rosa	16	0	0
Zweite Rosa	5	0	0
Harte-Flinze-Lagen	18	0	0
Total	51	0	3

Fig. 20. The positions of bivalve valves from the Mörnsheim Formation. The most common position was convex-up, this is conclusive evidence that the studied section was subject to weak current activity. Redrawn from Wilkin (2018).

One of the most common bivalves is the species *Liostrea socialis*. *Liostrea* lived attached to objects such as seaweed, driftwood and mobile cephalopods. They likely drifted into the lagoons by being attached to ammonite shells (Figure 21). As the weight of the bivalves increased, the ammonites sank to the inhospitable bottom waters killing both itself and its host. Some specimens could reach quite substantial sizes (up to 70 cm), meaning it must have been attached for some time (Barthel *et al.*, 1990).

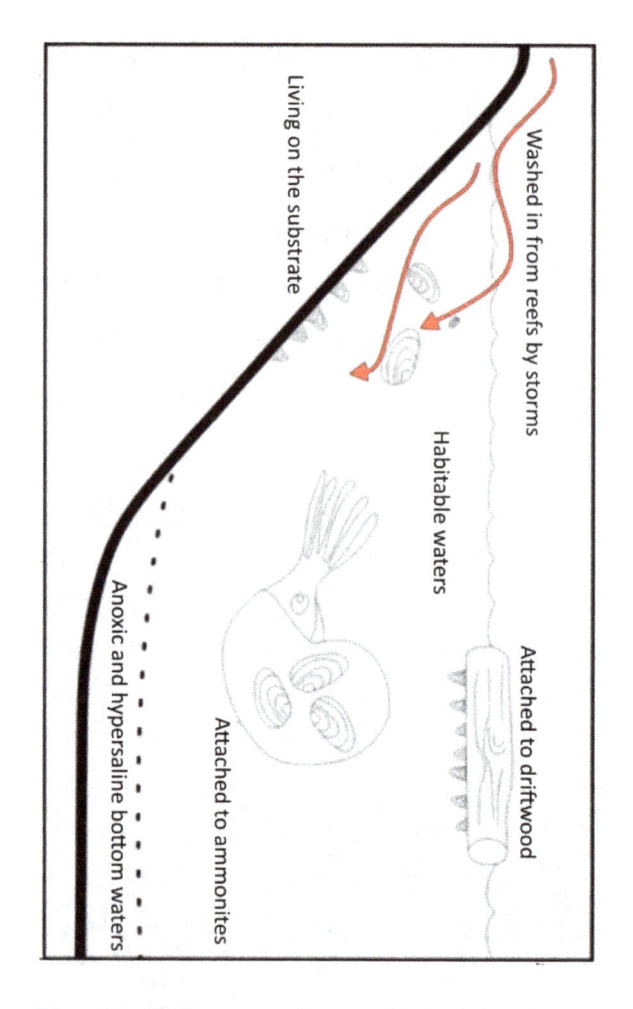

Fig. 21. Palaeoecology of bivalves in the Plattenkalks. Authors own work.

In the Mörnsheim Formation, the most common genera found during my excavation of the site in the summer of 2016 was the genus *Solemya* which can tell us about the palaeoenvironment of the basin. *Solemya* lived in shallow marine sulphide-rich environments which are evidence of anoxia (Stewart and Cavanaugh, 2006). *Solemya* was chemoautotrophic, using symbiotic bacteria living in specialised gill bacteriocytes to fix CO_2 via the Calvin Cycle with energy obtained from the oxidation of reduced sulphur compounds.

Cephalopods

Entirely marine cephalopods are the most advanced group of molluscs with highly developed brains and eyes. The eyes of molluscs evolved convergently with the eyes of vertebrates. Modern forms include *Nautilus*, the argonauts, squids and octopus. Classifying cephalopods is difficult. The scheme I will be using is Dzik's (1984) classification that divides Cephalopoda into three subclasses: Nautiloidea, Ammonidea and Coleoidea (squids, octopus and belemnites. Although nautiloids are known from the Plattenkalks, I will only be focusing on the ammonoids and coleoids in this chapter.

Ammonites are the most common fossils from the Plattenkalks. Jurassic ammonites are planispiral, coiled, forms and are two-dimensionally persevered (Figure 22). Sometimes, the siphuncle is preserved as a thin black line because of the original phosphatic makeup (Barthel *et al.*, 1990). In some examples, the suture lines are preserved (Figure 23). Ammonites hold a significant role in palaeontology as they are often used as zone fossils for biostratigraphy. Ammonite researchers have largely ignored Plattenkalks until recent years due to their compressed preservation (Schweigert, 2007; Schweigert, 2009).

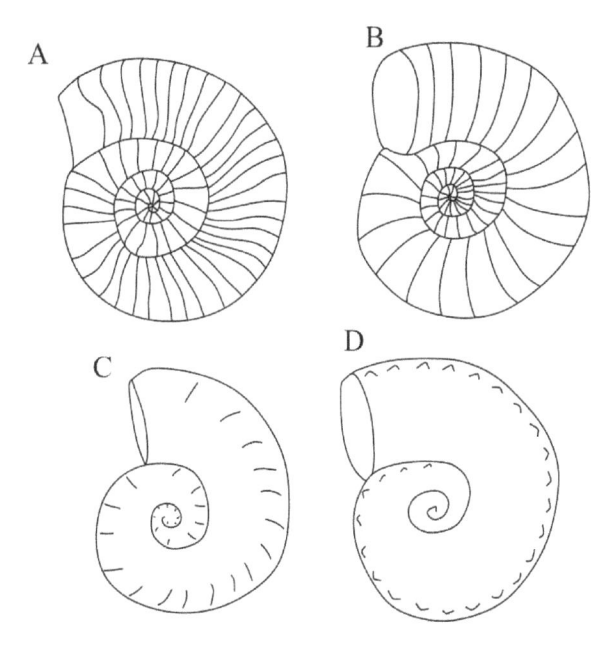

Fig. 22. Some common Ammonites recovered from the Plattenkalks. Authors own work.

A: *Subplanites*.

B: *Perisphinctes*.

C: *Neochetoceras steraspis.*

D: *Fontannesiella.*

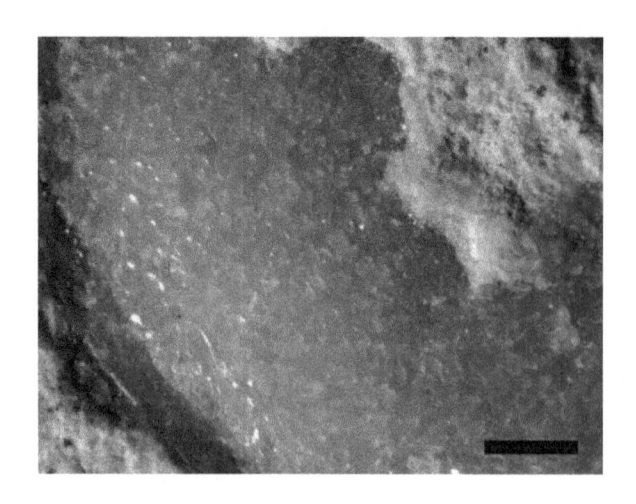

Fig. 23. Light microscope photograph of *Fontannesiella* from the Mörnsheim Formation clearly showing the ammonitic suture patterns. Scale bar= 2 mm. Authors own work.

Aptychi, the calcite coverings of the outer surface of ammonite jaws, are relatively common from the basins. Aptychi are commonly found apart from the rest of the ammonite shell and are rarely found in association with the rest of the ammonite (Figure 24). This would be expected if the ammonite were buoyant within the water column for some time after death. As the soft tissues around the head region rotted, the aptychi would have fallen out and were deposited on the sediment (Barthel *et al.*, 1990). The aptychi are the calcite coverings of the outer surface of ammonite jaws. Their primary function was acting as jaws but was secondarily adapted to work as opercula as a way of closing of the aperture to protect the soft tissues (Lehmann and Kulicki, 1990). Statically mapping of double valve aptychi relative to associated ammonites in Plattenkalks fully supports Lenmann's interpretation of aptychi being homologues of lower jaws (Seilacher, 1993). Ammonites feed on copepods, krill, and other small organisms (Kruta *et al.*, 2011).

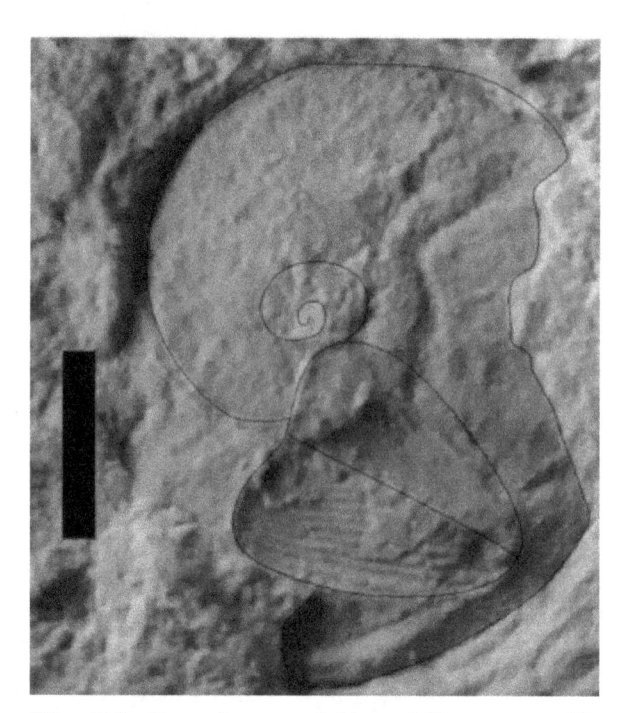

Fig. 24. Specimen of *Neochitoceras* with preserved aptychi, Mörnsheim Formation, near Mühlheim. Scale bar= 1 cm. Authors own work.

Belemnites are a now-extinct order of coleoids that thrived during the Mesozoic. Belemnites superficially resembled squids but had a heavy internal skeleton called a rostrum to serve as a counterbalance to the head and tentacles. Belemnites had ink sacs and ten arms each lined with 30-50 hooks called onychites to capture prey (belemnite paleobiology is summarised by Hoffmann and Stevens, 2020) (Figure 25).

Belemnites are often assumed to have been fast epipelagic swimmers based primarily on their superficial resemblance to extant teuthids. Variation in rostra shape was likely the result of different ecological niches and possibly water depths. Genera with short, thick guards may have been nektobenthic and slender, elongate forms being epipelagic swimmers (Rexfort and Mutterlose, 2009).

The most common genus of belemnite from the South German Jurassic was *Hibolites.* Juveniles are the most abundant belemnite fossils found within the Plattenkalk. There are two possible explanations for this. The first is that it represents a taphonomic bias, the larger adults were stronger swimmers and were, therefore, able to avoid being swept into the lagoon. If this explanation is correct, then it implies that belemnites were stronger swimmers than ammonites, as adult ammonites are commonly found in the plattenkalk. The second explanation is that juveniles were numerically more abundant either by being more common than adults in a general sense or the sheltered waters in or surrounding the basins served as a nursery (Wilkin, 2020). Such behaviour is observed in living cephalopods on shallow marine shelves and in areas of upwelling (Garofalo *et al.*, 2010). Both oceanographic features were present on the Franconian-Alb Platform during the Late Jurassic (Leinfelder, 1993).

Observations by Stevens *et al.* (2014) from the Late Jurassic Nusplingen Plattenkalk of Bavaria found that belemnite rostra were common and not transported into the basins by storms but rather lived in the lagoons their entire lives. Such a conclusion contrasts with the Solnhofen Plattenkalk, which was highly inhospitable to life, were belemnite rostra are rare and are thought to have been washed into the Plattenkalk post-mortem.

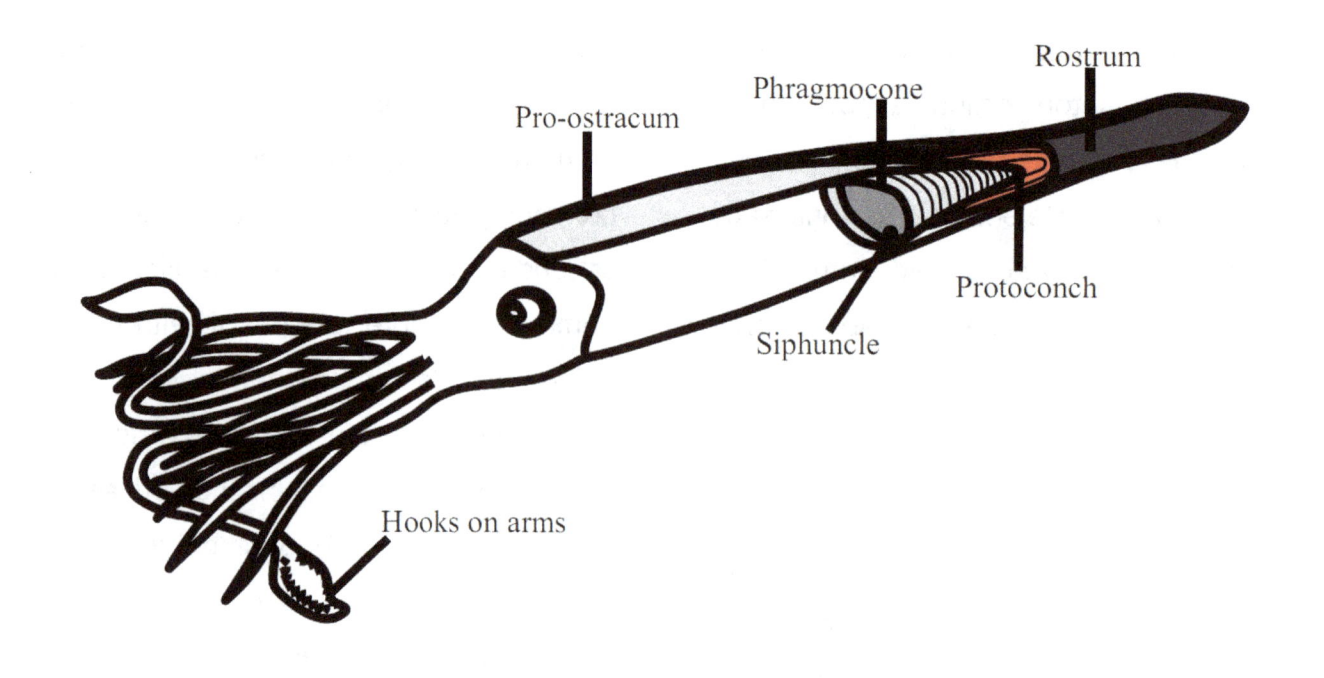

Fig. 25. Diagram showing belemnite anatomy. Author's own work.

Belemnites, due to their high Mg-Calcite shells, are relativity resistant to diagenetic alteration, making them ideal in palaeoclimate research. Regardless, great care must be taken when extracting isotopes for study. The quality and therefore reliability of the palaeoenvironmental reconstructions are highly dependent on the degree of alteration by post-depositional processes. Such diagenetic changes can modify the chemical, isotopic, and structural composition of fossil calcite. Oxygen isotopic measurements from the Late Jurassic Solnhofen Formation showed a mean surface water temperature of 26°C (Engust, 1961).

Other coleoids from the basins include vampire squids— Vampyromorpha— such as *Leptoteuthis gigas*. Vampire squids are known primarily from its gladius—the hard internal body part found in many cephalopods. Though in many examples from the Plattenkalks preserve soft tissues such as the mantel and tentacles (Fuchs, 2007; Sutton *et al.*, 2015) (Figure 26).

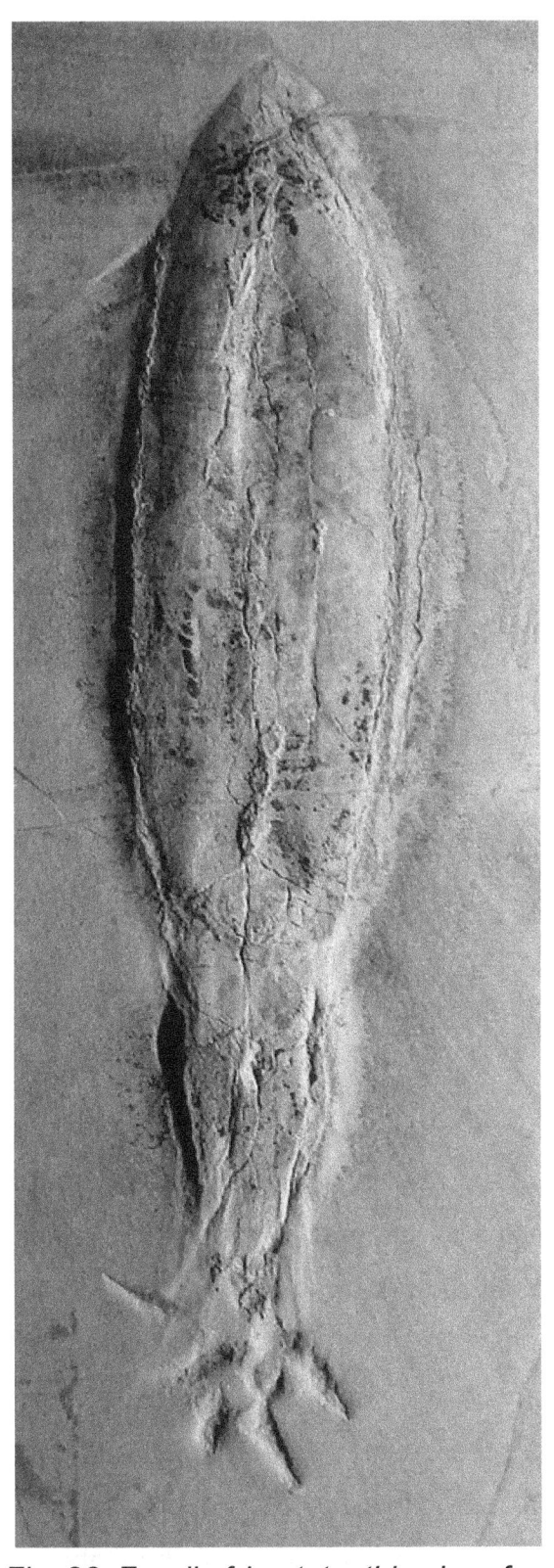

Fig. 26. Fossil of *Leptoteuthis gigas* from the Jura Museum. Photo courtesy of Wikimedia user Ghedoghedo.

Echinoderms

Echinoderms are a phylum of entirely marine invertebrate with a skeleton made of porous calcite plates held together by a thin layer of soft tissues making the skeleton mesodermal. After death, the soft tissues decay very quickly, causing the plates to become disarticulated.

The skeletons of echinoderms have five rayed-or pentamersal-symmetry, although some groups of sea urchins have subsequently evolved bilateral symmetry. Another essential feature of this group is the water-vascular system: a complex internal hydraulic system of tubes and bladders with extensions called tube-feet (or podia) that emerge through paired pores called ambulacra skeleton to the outside. Tube-feet serve a variety of purposes principally locomotion, respiration and feeding (Clarkson, 1993). All of the main groups of echinoderms are known from the Plattenkalks (Barthel *et al.*, 1990).

Sea-urchins (Echinoidea) inhabit a wide range of different biotopes from shallow marine waters to abyssal plain (Pawson, 1982; Miller and Pawson,

1989). Their basic shape is that of a globular test which is covered in spines (Figure 27), though these spines are reduced in sand dollars. The spines of echinoid are articulated on a ball and socket joint which allows them to move. After death, these spines fall away and disperse (Kidwell and Baumiller, 1990). However, in the Plattenkalk, echinoids spines can still be found attached in the animal (Barthel *et al.*, 1990). The echinoids found from southern Germany are mostly regular forms with nearly perfect pentameral (five-part) symmetry. They fed by grazing the reefs that surrounded the basin using their powerful self-sharpening jaws apparatus known as Aristotle's lantern to bite off bits of coral or graze algae much like modern forms today (Clarkson, 1993).

Crinoids are common in the geological record with most fossils being of the attached, sessile forms (so called "sea lilies"). A stalk comprised of five-sided ossicle held together by a soft tissue tube running through the centre makes up the main body of the animal and supports the cup-shaped calyx which contains the mouth and internal organs as well as five arm-like ambulacra which process hair-like filaments called pinnules which are used in filter-feeding. The only sessile crinoid from the Plattenkalks is *Millericrinus* (Schweigert *et al.*, 2008).

Fig. 27. *Rhabdocidaris orbignyana*. Redrawn from Grawe-Baumeister *et al.* (2000).

The most common crinoid and one of the most common fossils from the basins, in general, is the floating crinoid genus *Saccocoma* (Figure 28). However, a study by Milsom (1994) suggested that *Saccocoma* was benthic. *Saccocoma* can be used as an environmental indicator. If the arms are outstretched, then the environment was habitable. If the arms are contacted, then the environment was inhospitable either by hypersalinity and lack of oxygen (Matyszkiewicz, 1996; Martill, D. pers. comm, 2016).

Fig. 28. The floating crinoid *Saccocoma alpine*. Image courtesy of Wikimedia user Smokeybjb.

Fish

The most common of the Plattenkalk vertebrates are fish. The Plattenkalks preserve a remarkable diversity of fishes, including sharks, rays, and ratfish. However, it is teleost's, bony fishes that are by far the most abundant. Fish are also incredibly useful as palaeoenvironmental indicators because they are comprised of many elements makes them extremely sensitive to different depositional conditions (see Pan, 2019; Wilkin, 2019a for a review of fish taphonomy from the Solnhofen and Mörnsheim formations respectively).

One of the more well-known fishes from the Plattenkalks is *Caturus*. *Caturus* was a fast-epipelagic predatory fish equipped with light scales and symmetrical caudal fins, allowing for rapid movement while pursuing prey (Müller, 2011). More direct evidence for its predatory lifestyle can be found in a fossil of *Caturus* with another fish in its mouth or stomach. Still, other fish show evidence of predation with bites taken out of them. Yet, others chocked while eating a meal too big for them (Figure 29).

The most common genus is the early teleostean (modern bony fish) *Leptolepides*. This sprat-like form is commonly found in mass mortality assemblages. Mass mortality assemblages are identifiable by several features such as gaping jaws, hyperextended branchiostegal rays, and splayed fins (Pan *et al.*, 2019; Wilkin, 2019a).

Some fishes are so well preserved that details such as colour patterns are still present. One example is a *Thrissops* from the Ettling Plattenkalk described by Tischlinger (1998) (Figure 30).

Thrissops, at 55cm, was a reasonably sizeable predatory fish which preyed on other smaller fishes such as *Orthogonikleithrus hoelli* as stomach contents (Ebert *et al.*, 2015).

Fig. 29. *Ebertichthys ettlingensis* eating *Orthogonikleithrus*. JME. Authors own work.

Fig. 30. One of the most stunning fossils ever discovered. *Thrissops* with preserved patterning. JME. Authors own work.

Reptiles

The Mesozoic is often referred to as the "Age of Reptiles". The genera *Eichstaettisaurus* (Figure 31), *Ardeosaurus*, *Palaeolacerta*, and *Bavarisaurus* from the Solhnofen Formation represent some of the oldest known articulated squamates (Barthel *et al.*, 1990). Such fossils, especially *Eichstaettisaurus*, are useful in understanding early squamate evolution. *Eichstaettisaurus* is a gekkonomorph lizard, closely related to geckos and is known from the Late Jurassic to the Early Cretaceous of Germany and Italy (Evens *et al.*, 2004; Simões *et al.*, 2017).

Both terrestrial and aquatic crocodilians are also known the most common being *Alligatorellus beaumonti* (Figure 32). *Alligatorellus* is a terrestrial crocodile that was washed into the basins from the surrounding islands and is known from the Plattenkalk at Kelheim (Schwarz-Wings *et al.*, 2011) and also from the French Plattenkalk of Cerin (Gervais, 1871).

The Plattenkalks also contains rare examples of ichthyosaurs. Ichthyosaurs are the group of marine reptiles most perfectly adapted for an aquatic lifestyle having streamlined fish-shaped bodies. The most common ichthyosaur from the Plattenkalks is *Aegirosaurus leptospondylus*. Ichthyosaurs swam using what is known as thunniform locomotion, that is as a type of axial movement that involves moving its body and tail from side-to-side-to while using their paddles for stabilisation. It was a small species, with adults being less than two meters in length (Bardet and Fernández, 2000).

The south German limestones also preserve a diverse range of Mesozoic turtles with seven genera known thus far (e.g. von Meyer, 1839, 1864; Wellnhofer, 1967; Gaffney, 1975; Joyce, 2000, 2003. Turtles are shelled reptiles belonging to the order Testudines. Fossil turtles are quite rare in the fossil record before the Cenozoic, and the group's origins remain somewhat of an enigma. Still, they are thought to have evolved during the Middle Jurassic (Joyce, 2017). The turtles from the Plattenkalks are all marine species with relativity long tails - a primitive trait lost in modern forms — and even juveniles — are known (Figure 33).

Fig. 31. Skeleton of *Eichstaettisaurus*. JME. Authors own work.

Fig. 32. The crocodilian *Alligatorellus* sp. JME. Authors own work.

Fig. 33. The sea turtle *Eurysternum*. JME. Authors own work.

Pterosaurs were flying reptiles that lived during the Mesozoic. Some species could reach truly gigantic sizes, up to 10 metres from wing-tip-to-wing-tip or greater. However, those from the Solnhofen are much more modest in size with the largest species, *Rhamphorynchus muensteri*, having a 1.8-metre wingspan (Wellnhofer, 1975). What these pterosaurs lacked in size they more than made up for in their beautifully detailed preservation. Not only are complete, articulated skeletons commonly found but soft tissues such as wing membranes and even pycnofibers can be preserved (Wellnhofer, 1970; Bennett, 2002; Frey *et al.*, 2003; Elgin *et al.*, 2011; Vidovic and Martill, 2014).

The first documented pterosaur fossil was discovered sometime between 1767 and 1784 from the Jurassic Solnhofen Limestones of Southern Germany. The first person to ever describe a pterosaur was an Italian historian and Voltaire's secretary Cosimo Alessandro Collini in 1784. This animal would later be named *Pterodactylus*.

The fossil was first thought to be that of an amphibious animal (Wagler, 1830). The fossil was then passed from scientist to scientist until coming into the precession of a French anatomist George Curvier who realised in an 1801 paper that the skeleton belonged to a flying reptile and later in 1809 coined the term pterodactyl meaning winged finger.

It would take a few years for the idea of flying reptiles to be entirely accepted by other scientists, and some, most notably English anatomist Edward Newman (1843), even believed, incorrectly, that pterosaurs were a type of flying marsupial. The history of pterosaur research is a long and fascinating one, and if you are interested and would recommend Wellnofer (2009a) and

Witton (2013) also has some excellent information.

Pterodactylus is one of the most common pterosaurs, and the abundance of well-preserved specimens has allowed researchers to reliably reconstruct not just the skeletal but also the soft tissue anatomy. *Pterodactylus* was modestly sized with a wingspan of up to 1-meter. A low midline crest grew from the skull when the animal reached sexual maturity and was likely used for display (Bennett, 2012). Due to elongated vertebrae, the neck was much longer than that of its relatives. *Pterodactylus* had a small throat pouch that was used to transport a small amount of food; perhaps this is evidence of parental care. A closely related genus was *Aerodactylus* (Figure 34).

Rhamphorhynchus is one of the best-known pterosaurs and was the Jurassic equivalent of a seagull with a wingspan of 1.8 metres. The wing membranes of these pterosaurs are often well preserved (Figure 35). Their jaws housed needle-like forward pointed teeth, perfect for catching slippery fish. Fishbones are commonly found in the stomach contents of these animals (Wellnhofer, 1991; Tischlinger, 2010; Frey and Tischlinger, 2012). *Rhamphorhynchus* dip feed – that is swiping fish and squids (see Hoffmann *et al*, 2020) close to the water surface while on the wing (Chatterjee and Templin, 2004; Humphries *et* al., 2007; Witton, 2008, 2013, 2018;; Hoffmann *et al.*, 2020)or when floating in on the water surface (Hone and Henderson, 2014). Sometimes the tables would turn, and the *Rhamphorhynchus* ended up being dinner for the fish. This is shown by a remarkable fossil from the Solnhofen of Germany, which shows a *Rhamphorhynchus* in the jaws of a large fish called *Aspidorhynchus* (Frey and Tischlinger, 2012) (Figure 36).

(Page 46) Fig. 34. *Aerodactylus* skeleton. Image courtesy of Steve Vidovic and David Martill.

(Page 47) Fig. 35. Rhamphorhynchus showing wing membrane. JEM. Authors own work.

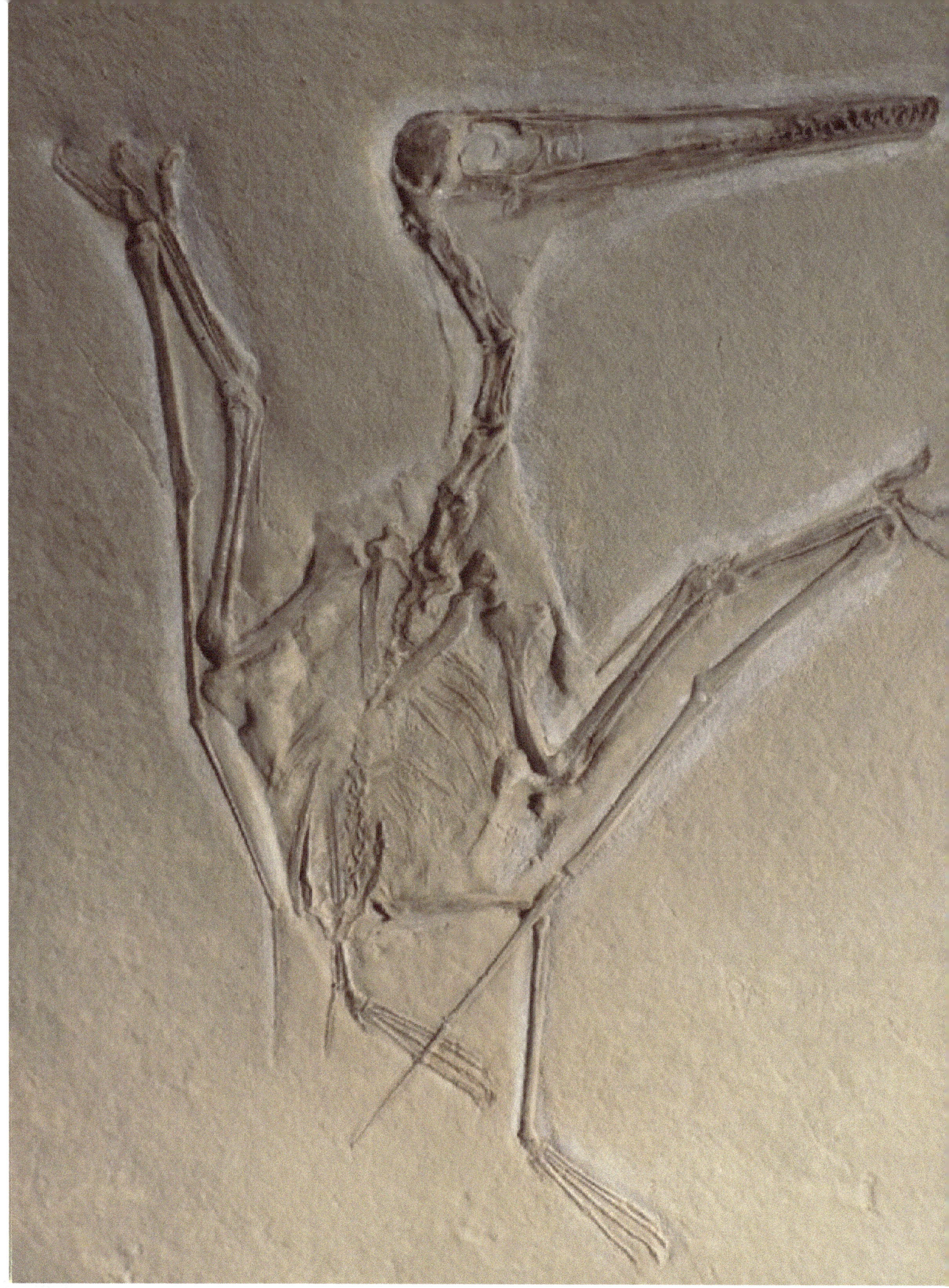

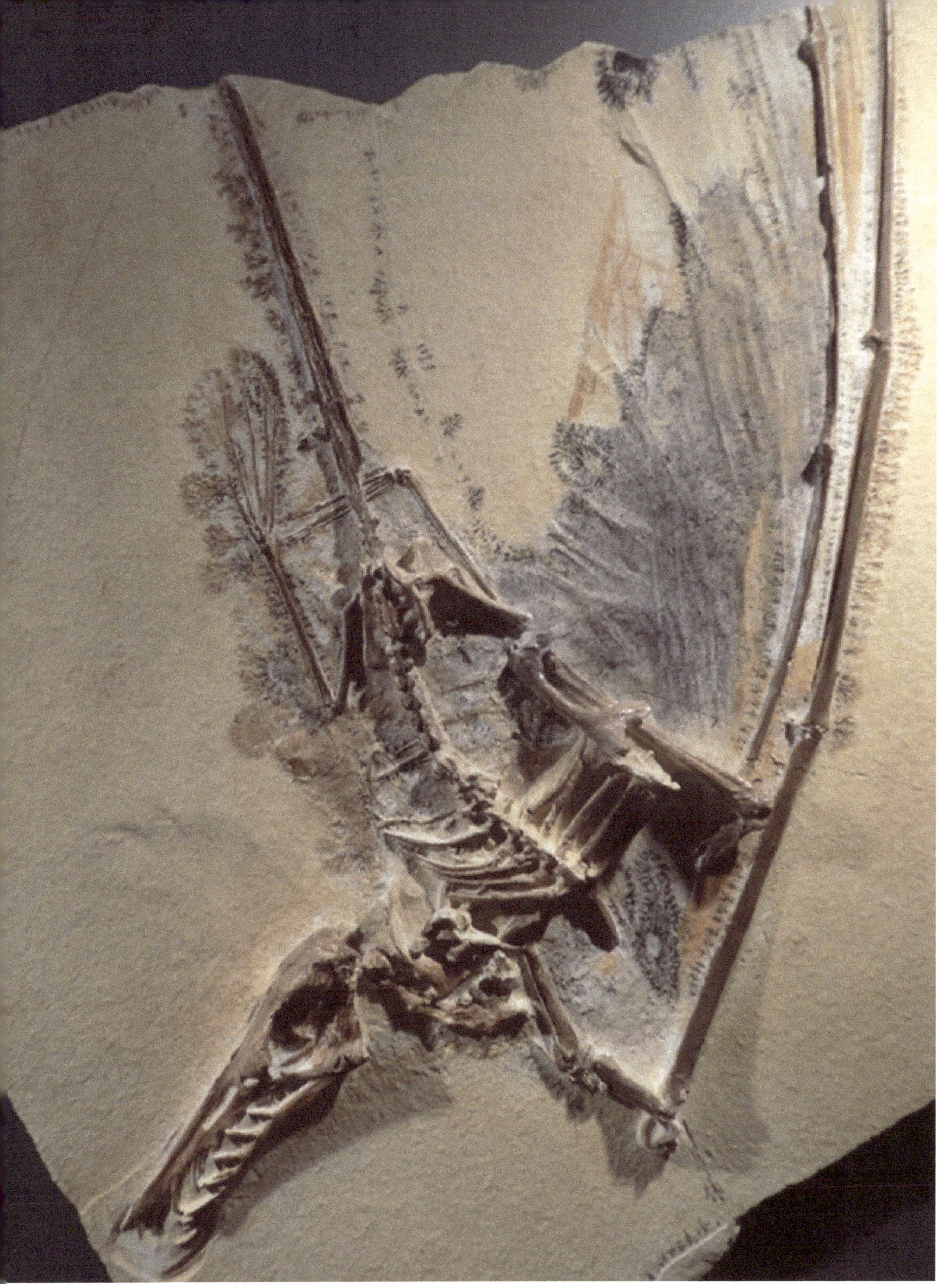

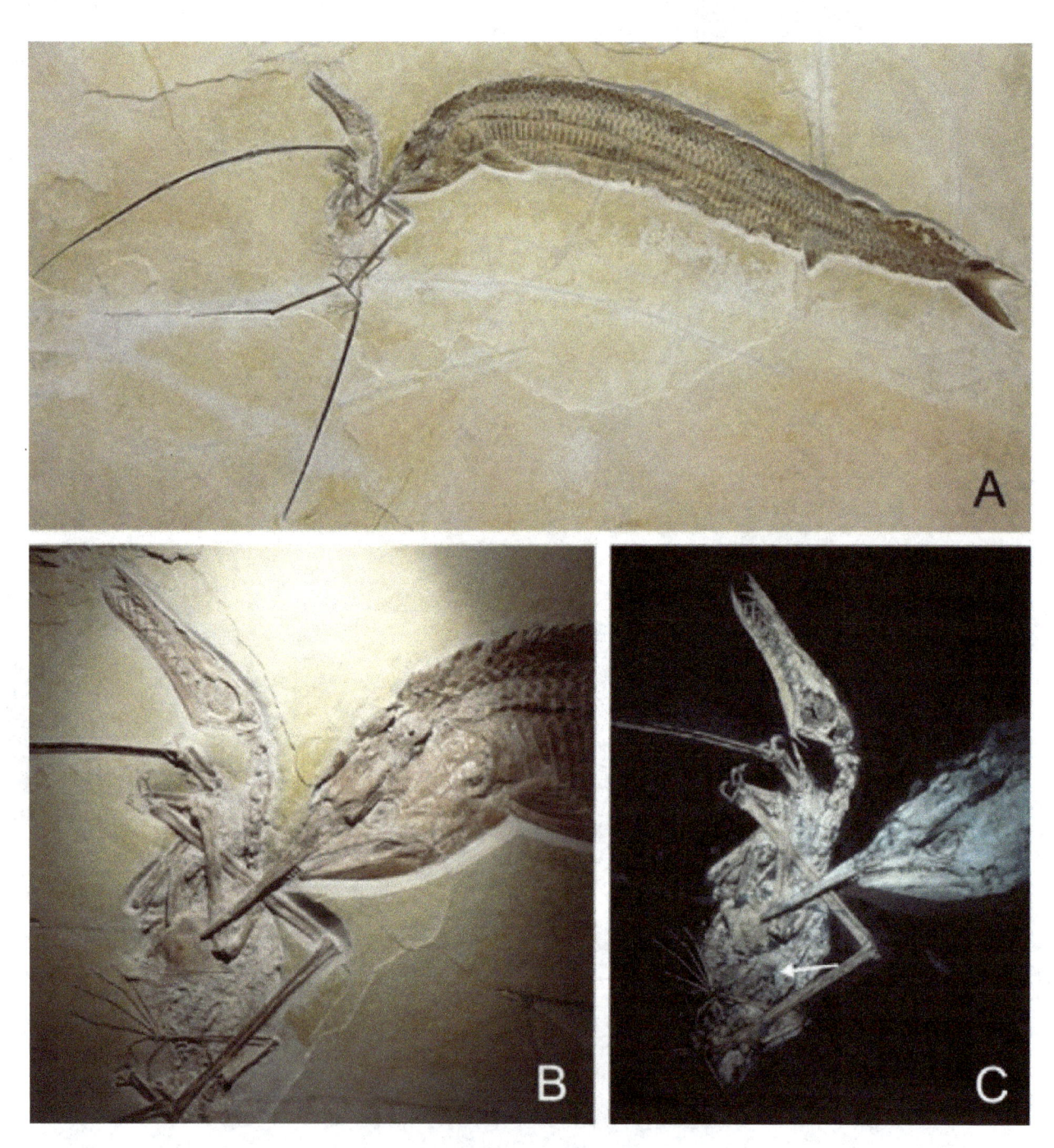

Fig. 36. *Rhamphorhynchus* (specimen WDC CSG 255) with a *Leptolepides* fish trapped in the pharynx and caught in the jaws of an *Aspidorhynchus*. Image courtesy of *PLoS ONE*, Eberhard "Dino" Frey and Helmut Tischlinger.

Ctenochasma had hundreds of bristle-like teeth that projected outwards forming a large spoon-shaped basket (Wellnhofer, 1970; Jouve, 2004). *Ctenochasma* was a filter-feeder that used its teeth to filter invertebrates out of the mud (Wellnhofer, 1991). Its long hindlimbs and broad feet indicate that *Ctenochasma* spent a significant amount of its time wading in shallow water searching for prey. Closely related and sharing the same bizarre snout design and feeding style was *Gnathosaurus subulatus* (Wellnhofer, 1970, 1978, 1991; Fastnacht, 2005)

Anurognathus was a small pterosaur with a body only 9 cm long and a wingspan of 35-50 cm. They have short tails and faces with small needle-like teeth. They were very agile as to catch flying insects, like a modern nightjar (Döderlein, 1923; Unwin, 2005; Bennett, 2007; Witton, 2013), and their jaw musculature was adapted for very rapid closure (Ősi, 2010). Anurognathians had large eyes, suggesting they may have been nocturnal (Bennett, 2007).

Dinosaurs

Dinosaur remains are rare in the Plattenkalks, with only four genera known: the *Compsognathus*, *Juravenator*, *Ostromia* and *Sciurumimus*, all of which are theropods. For 160 million years, dinosaurs were the dominant land animals and are represented today by over 10,000 living species of bird. They are a varied group of animals from a taxonomic, morphological and ecological standpoints. Like modern birds and crocodiles, dinosaurs laid eggs and took care of their young. Although many were small, Dinosauria contained the largest terrestrial organisms the world has ever seen. The sauropod *Argentinosaurus*, for example, was one of the largest-known land animals of all time, if not the largest, with length estimates reaching 30 meters (Appenzeller, 1994; Carpenter, 2006) and weight estimates given by Mazzetta *et al.* (2004) of 60–88 tonnes (66–97 short tons).

Dinosauria is a clade of archosaurs that first evolved around 240 million years ago during Middle Triassic. The oldest undisputed dinosaur remains are 231 million years old from South

America and include the likes of *Eoraptor* (Alcober and Martinez, 2010) and *Herrerasaurus*. The oldest dinosaurs known from Germany are known from the Upper Triassic and includes the protosauropod *Plateosaurus* (von Meyer, 1837; von Huene, 1926).

Dinosaurs are split into two groups based on the shape of the hips: the lizards-hipped saurischians and the bird-hipped ornithischians. The saurischians, ironically the group that contains birds, are further split into the sauropods – giant long-necks herbivores – and theropods which contained mostly carnivorous forms. Theropod diversity is summarised by Hendrickx *et al.* (2015).

Compsognathus was a small, lightly built theropod with a relativity long neck (Figure 37). *Compsognathus* had asymmetrical hands with a short thumb, and the outermost finger was elongated. Preserved stomach shows that they fed on small lizards and mammals. *Compsognathus* was originally described by German palaeontologist Johann Wagner (1859) giving it the name *C. longipes* and described the specimen in more detail in 1861. Thomas Huxley (1868) noticed the close similarities between *Archaeopteryx* and *Compsognathus,* which further proved the dinosaurian origin of birds. A second specimen was discovered from the Tithonian Canjuers Plattenkalk of Provence in France in 1971 (Bidar *et al.*, 1972). Relatives of *Compsognathus* have been found in China (Ji *et al.*, 2007) which, also given that *Juraventor* was also feathered it is highly likely that *Compsognathus* was also feathered (Peyer, 2006).

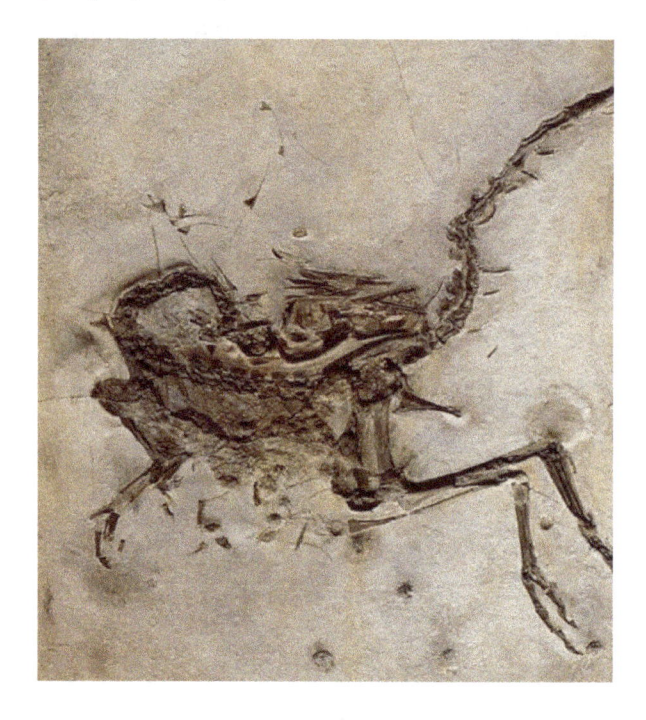

Fig. 37. *Compsognathus* cast. Image courtesy of Wikimedia user MatthiasKabel.

Juravenator was a small, 75 centimetre, predator known only from a single juvenile specimen from the

Kimmeridgian Painten Formation near Eichstätt. The specimen was discovered in 1999 but was not formally described until 2006 by Ursula Göhlich and Luis Chiappe. Small, Late Jurassic carnivorous dinosaurs are rare worldwide, and in Europe are represented by just two poorly preserved *Compsognathus* skeletons. The fossil preserves evidence of primitive filament-like feathers. The size of *Juravenator*'s scleral rings around the orbit suggests that it may have been nocturnal (Schmitz and Motani, 2011) (Figure 38).

Ostromia is a very bird-like dinosaur from the Painten Formation. For years it was thought to have been a specimen of *Archaeopteryx* named the Haarlem Specimen (Ostrom, 1970). A re-examination by Foth and Rauhut (2017) found no diagnostic features for *Archaeopteryx*. The study did show morphological characteristics with anchiornithids making *Ostromia* the only member of this group known from outside the Tiaojushan Formation of China. Such morphological characteristics includes the shape of the furrows on the finger bones (Figure 39) and the public shaft flexes backwards in *Ostromia* much in the same way as *Anchiornis* as opposed to *Archaeopteryx* (Foth and Rauhut, 2017).

Fig. 38. *Juravenator*. JME. Authors own work.

The final dinosaur from the Plattenkalks is *Sciurumimus* from the Painten Formation (Figure 40). *Sciurumimus* is interesting as soft tissues are preserved in several areas of the skeleton. Most of these soft tissues represent integumentary structures (skins and protofeathers), except for a short section of fossilized tissue along the posterior edge of the tibia, which may possibly be muscle tissue (Rauhut *et al.*, 2012).

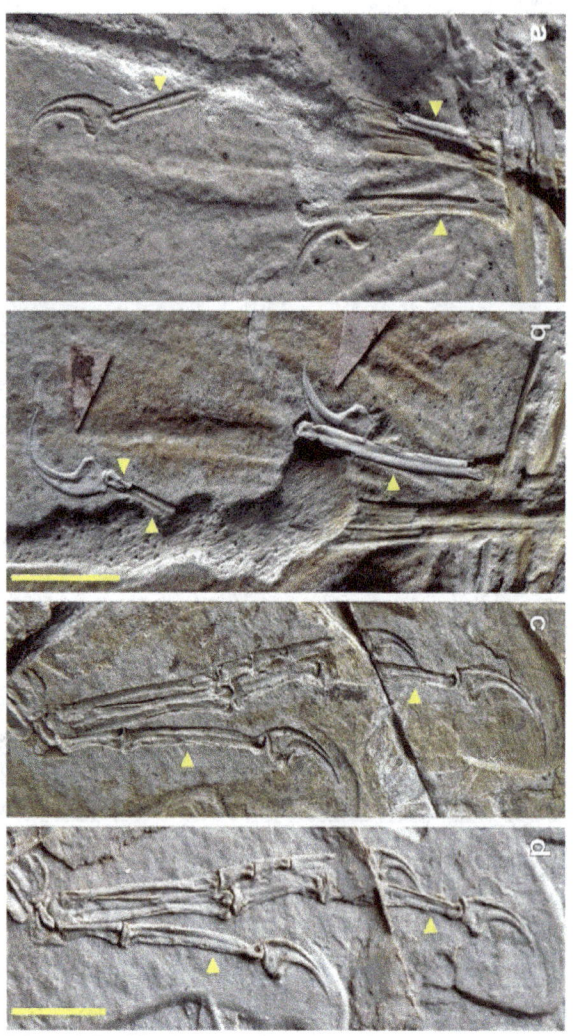

Fig. 39. Furrows (yellow arrows) in the hand bones of *Ostromia* (a,b) compared to China's *Anchiornis* (c,d). WikiMedia CC.4.0

The best soft tissues are found on the tail, which includes large patches of skin and very fine, long, hair-like monofilaments protofeathers. The skin, unlike *Juravenator's*, is smooth with in sign of scales (Rauhut *et al.*, 2012; Foth *et al.*, 2020). *Sciurumimus* was originally classified as a basal megalosauroid (Rauhut *et al.*, 2012), making it the only definite rotofeathers known in theropods were in primitive coelurosaurs showing that small megalosaurs, at least juvenile forms, where fuzzy. Godefroit *et al.* (2013), however, disagreed with Rauhut *et al.* (2012) instead suggesting that *Sciurumimus* as a basal coelurosaur, which would mean that feathers were limited to this clade. Still, Foth *et al.* (2020, pp. 36-37) proposed that *Sciurumimus'* coelurosaurian characters are in fact due to the only known specimen was a very young individual as heterochrony seems to have played an important role in the evolution of coelurosaurian theropods (e.g. Bhullar *et al.* 2012; Foth *et al.* 2016), and therefore a basal tetanuran placement for *Sciurumimus* is more likely

Fig. 40. *Sciurumimus*. JME. Authors own work

Archaeopteryx

Birds are the most species-rich group of living tetrapods with the early evolution of birds being at the forefront of palaeontological research. The family Archaeopterygidae are a group of maniraptoran dinosaurs that lived during the Late Jurassic period of Central Europe. The family only contains the two genera: *Archaeopteryx* and, somewhat controversy, *Wellnhoferia*.

Archaeopteryx sometimes referred to as *Urvogal* meaning "original bird" in German, is a textbook example of a transitional fossil and a good candidate for the ancestor of modern birds. It had avian features such as wings and feathers but also reptilian features like a long bony tail, claws on its hands and teeth (Figure 41). The idea that birds evolved from dinosaurs is an old one, and the theory can be traced to Thomas Huxley in the 1860s and 1870s after a series of papers published following the discovery of *Archaeopteryx* and, to a lesser extent, *Compsognathus* (Huxley, 1868a; 1868b; 1870).

Archaeopteryx was named from a single feather in 1861 by Von Meyer (1861; 1862- see also Griffiths, 1996 for an analysis of this specimen). However, a paper published in *Nature* by Thomas Kaye *et al.* (2019) sheds doubt on the avian origin of the feather. The single feather is missing the calamus, part of the quill that inserts into a follicle in the skin, but by using Laser-Stimulated Fluorescence (LSF) can detect the geochemical signatures of missing elements. From this, the researchers concluded that the feather structure did not match that of *Archaeopteryx* leading to the possibility that a yet unknown bird or bird-like dinosaur lived in Central Europe at the same time as *Archaeopteryx* and *Wellnhoferia*.

The preservation in the Plattenkalks is so good that melanosomes, organelles responsible for pigmentation, are present. By comparing the shape of these structures under a Scanning Electron Microscope to those of modern birds, it is possible to work out the original colouration. Analysis of *Archaeopteryx* feathers by Carney *et al.* (2012) showed that *Archaeopteryx* was black.

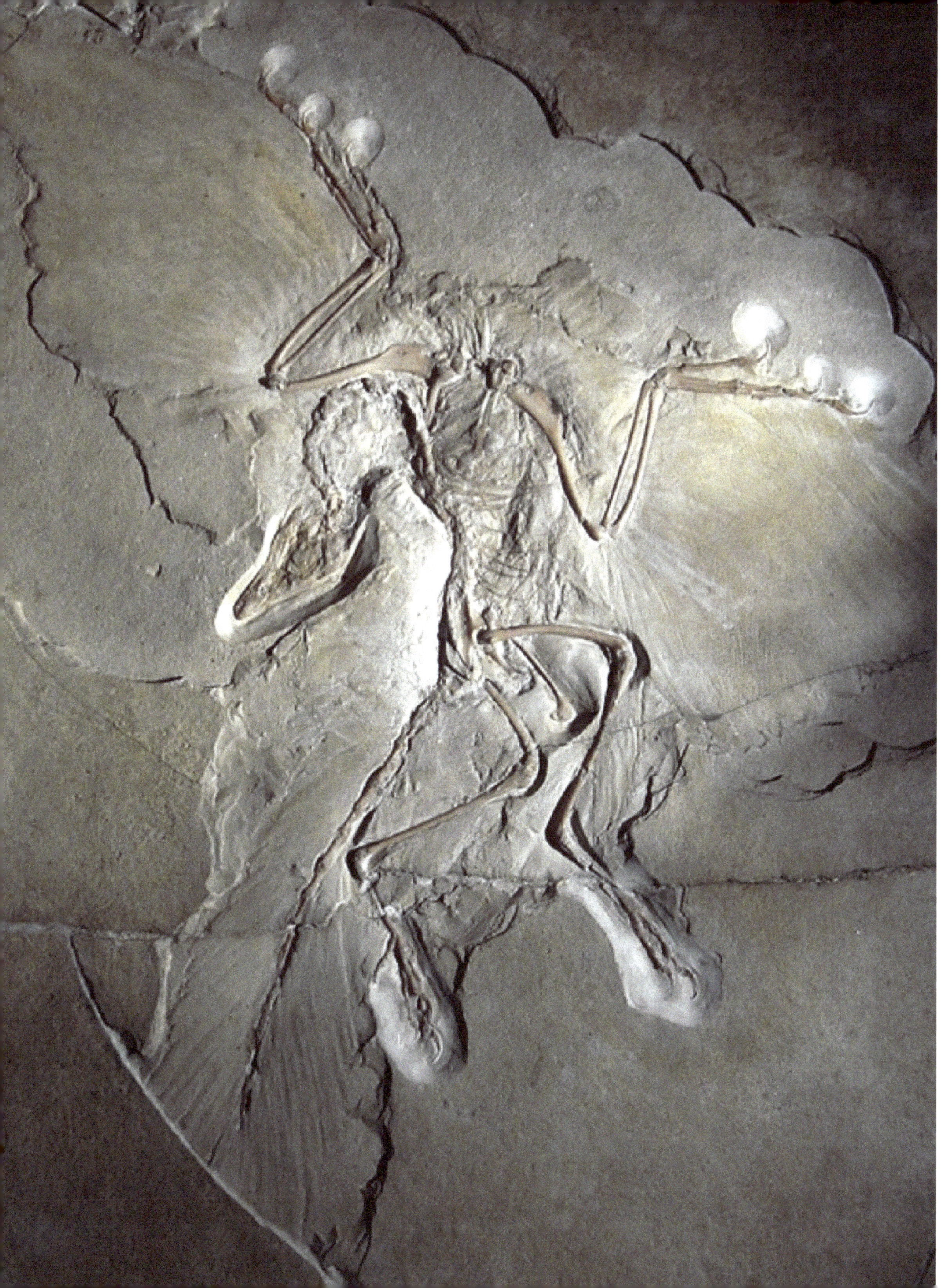

(Previous page) nFig. 41. Perhaps the most iconic fossil in the World- the Berlin Specimen of *Archaeopteryx*. Image courtesy of H. Raab.

Archaeopteryx was about the size of a magpie. None of the known specimens show stomach contents, but the shape of the teeth suggests a diet consisting of insects and small vertebrates. and fed on insects (Mayr, 2017, pp.21). The claws on the hands and feet suggest that *Archaeopteryx* may have been arboreal (Elzanowski, 2002) with the shape of the claws, which most are still perseverved with their horny sheaths, corresponds with those of tree climbing birds and mammals (Feduccia, 1993). However, this theory is questioned by Pike and Maitland (2006) who argued that mode of life could not be predicted with any certainty using measurements of either claw radius or claw angle and Wellnhofer (2009) contended that the hindlimb morphology of *Archaeopteryx* does not indicate a climbing habit and instead was predominantly terrestrial. This view is supported by other authors (e.g. Ostrom, 1976).

The question of whether *Archaeopteryx* was capable of flight is a topic covered, and often contradicted, by hundreds of scientific papers (e.g. Olson and Feduccia, 1979; Feduccia and Tordoff, 1979; Speakman and Thomson, 1994; Norberg, 1995; Chatterjee and Templin, 2003; Senter, 2006; Longrich, 2009; Nudds *et al.*, 2010; Voeten *et al.*, 2018 and many more) and would require a PhD's worth of work to cover in sufficient detail. The feathers were asymmetrical, implying that they were used to generate lift. The lack of a bony breastbone meant that *Archaeopteryx* could not fly in the same manner as modern birds. A paper by Voeten *et al.* (2018) suggests that *Archaeopteryx* could fly but used more shoulder action that modern birds require. Chatterjee and Templin (2003) suggested that the lack of powerful flight muscles and complex wing movements meant that *Archaeopteryx* was incapable of ground takeoff. Instead the authors suggested that *Archaeopteryx* may have made short flights between trees, utilizing a novel method of phugoid gliding.

Wellnhoferia was discovered near Eichstätt in the 1960s and was initially described as the largest specimen of *Archaeopteryx* by Peter Wellnhofer (1988). Although very similar to *Archaeopteryx* (Figure 42), hence why it was initially classified as one, *Wellnhoferia* had a reduced tail and fourth toe and a myriad of other features which led Polish palaeontologist Andrzej Elżanowski to assign a new genus and species in 2001. This reclassification has been supported by subsequent studies (Senter and Robins, 2003).

Fig. 42. *Wellnhoferia*. Image courtesy of Wikimedia user FerdiBf.

Final Thoughts

The south German Plattenkalks contain the exceptionally well persevered bodies of both marine and terrestrial organisms that lived in central Europe during the Late Jurassic. The Plattenkalks formed in the shallow warm waters on the north western margin of the Tethys Ocean within marine basins separated from the turbinate waters of the wider ocean by coral and sponge reefs. The basins allowed for calm conditions which in tow resulted in less water mixing and unhospitable conditions for life. The lack of scavengers, bioturbation, wave action, and – in some of the basins – microbial mats allowed for a taxonomically varied and wonderfully preserved biota.

Fossils are somewhat difficult to find in the Plattenkalks, though each basin is different, a whole day of fieldwork very often leads to a small collection of ammonites and if you are lucky a disarticulated fish. The fossils that are recovered are some of the best preserved and most scientifically important in the world. Fossils, or at least replicas, from the Plattenkalks, are a staple of museums around the world. Some of the best collections of Solnhofen material are the Jura Museum Eichstätt, Bürgermeister-Müller-Museum in Solnhofen, Paläontologische Museum in Munich, and the Museum für Naturkunde in Berlin.

The south German Plattenkalks, although the most famous and well studied lithographic limestones in the world, are not unique. Other Plattenkalk localities dating to the Late Jurassic include the Owadów-Brzezinki quarry in central Poland (Kin and Błażejowski, 2012; Kin et al., 2012, 2013; Błażejowski et al., 2016), Cerin in south eastern France (Bausch et al., 1994; Enay et al., 1994; Gaillard et al., 1994; Wenz et al., 1994).

Fossils from these remarkable localities across southern Germany provide a snapshot of a world frozen in time. Not only are articulated skeletons – something rarely seen in the fossil record – somewhat common providing researchers with important insights in to how different animal groups evolved – this is especially true with *Archaeopteryx* – but behaviours such as "death-marches" are also known.

Bibliography

Alcober, O. A. & Martinez, R.N. (2010). A new herrerasaurid (Dinosauria, Saurischia) from the Upper Triassic Ischigualasto Formation of northwestern Argentina. *ZooKeys* **63**: 55–81.

Appenzeller, T. (1994). Argentine dinos vie for heavyweight titles. *Science* **266** (5192): 1805.

Amiot, R., Lécuyer, C., Buffetaut, E., Escarguel, G., Fluteau, F., Martineau, F. (2006). Oxygen isotopes from biogenic apatites suggest widespread endothermy in Cretaceous dinosaurs. *Earth and Planetary Science Letters* **246**: 41–54.

Arratia, G. (2016). New remarkable Late Jurassic teleosts from southern Germany: Ascalaboidae n. fam., its content, morphology, and phylogenetic relationships. *Fossil Record* **19**:31-59.

Arratia, G., Schultze,H., Tischlinger,H., Viohl,G (eds.). (2015a). *Solnhofen: Ein fenster in die Jurazeit VOL. 1.* Munich: Verlag Dr. Friedrich Pfeil.

Arratia, G., Schultze,H., Tischlinger,H., Viohl,G (eds.). (2015b). *Solnhofen: Ein fenster in die Jurazeit VOL. 2.* Munich: Verlag Dr. Friedrich Pfeil.

Bakker, R.T. (1986). *The Dinosaur Heresies.* New York: William Morrow.

Bardet, N., Fernández, M. (2000). A new ichthyosaur from the Upper Jurassic lithographic limestones of Bavaria. *Journal of Paleontology.* **74** (3): 503–511.

Barthel, K.W., Swinburne, N.H.M., Conway Morris, S. (1990). *Solnhofen: A study in Mesozoic palaeontology.* Cambridge: University of Cambridge Press.

Baush, W.M., Viohl, G., Bernier, P., Barale, G., Bourseau, J-P., Buffetaut, E., \Gaillard, C., Gall, J-C., Wenz, S. (1994). Eichstätt and Cerin: geochemical comparison and definition of two different plattenkalk types. *Geobios* **27. (sup. 1)**: 107-125.

Bennett, S.C. (2002) Soft tissue preservation of the cranial crest of the pterosaur *Germanodactylus* from Solnhofen. *Journal of Vertebrate Paleontology* **22**: 43–48.

Bennett, S.C (2007). A second specimen of the pterosaur *Anurognathus ammoni. Paläontologische Zeitschrift* **81**: 376-398.

Bennett, S.C. (2012). New information on body size and cranial display structures of *Pterodactylus antiquus,* with a revision of the genus. *Paläontologische Zeitschrift* **87**: 269–289.

Bhullar, B-A.S., Marugán-Lobón, J., Racimo, F., Bever, G.S., Rowe, T.B., Norell, M.A., Abzhanov, A. (2012) Birds have paedomorphic dinosaur skulls. *Nature* **487**:223–226.

Bidar, A., Demay, L., Thomel G. (1972). *Compsognathus corallestris*, une nouvelle espèce de dinosaurien théropode du Portlandien de Canjuers (Sud-Est de la France). *Annales du*

Muséum d'Histoire Naturelle de Nice **1**: 9–40.

Bieńkowska-Wasiluk, M. (2010). Taphonomy of Oligocene teleost fishes from the Outer Carpathians of Poland. *Acta Geologica Polonica* **60**(4):479-533.

Błażejowski, B., Gieszcz, P., Tyborowski, D. (2016). New finds of well-preserved Tithonian (Late Jurassic) fossils from the Owadów-Brzezinki Quarry, central Poland: a review and perspectives. *Volumina Jurassica* **14**: 123–132.

Błażejowski, B., Lambers, P., Gieszcz, P., Tyborowski, D., Binkowski, M. (2015). Late Jurassic jaw bones of Halecomorph fish (Actinopterygii: Halecomorphi) studied with X-ray microcomputed tomography. *Palaeontologia Electronica* **18**.3(53A): 1-10.

Bradet, N., Fernandez, M.S. (2000). A new ichthyosaur from the Upper Jurassic lithographic limestones of Bavaria. *Journal of Paleontology* **74**(3): 503-511.

Brenchley, P.J., Harper, D.A.T. (1998). *Palaeoecology: Ecosystems, environments and evolution*. London: Chapman and Hall.

Carpenter, K. (2006). Biggest of the big: a critical re-evaluation of the mega-sauropod *Amphicoelias fragillimus* Cope, 1878. *Paleontology and Geology of the Upper Jurassic Morrison Formation. New Mexico Museum of Natural History and Science Bulletin* **36**: 131-138.

Carney, R; Vinther, J., Shawkey, M. D., d'Alba, L., Ackermann, J. (2012). New evidence on the colour and nature of the isolated Archaeopteryx feather. *Nature Communications* **3**: 637.

Charig, A.J., Greenaway, F., Milner, A.C., Walker, C.A., Whybrow, P.J. (1986). *Archaeopteryx* is not a forgery. *Science* **232**(4750): 622-6.

Chatterjee, S., Templin, R. J. (2003). The flight of *Archaeopteryx*. *Naturwissenschaften* **90** (1): 27-32.

Chatterjee, S., Templin, R. J. (2004). Posture, locomotion and palaeoecology of pterosaurs. *Geological Society of America Special Publications* **376**: 1-64.

Chen, P., Dong, Z., Zhen, S. (1998). An exceptionally well-preserved theropod dinosaur from the Yixian Formation of China. *Nature* **391** (8): 147–152.

Chiappe, L.M., Göhlich, U.B. (2010). Anatomy of *Juravenator starki* (Theropoda: Coelurosauria) from the Late Jurassic of Germany. *Neues Jahrbuch für Geologie und Paläontologie - Abhandlungen* **258** (3): 257–296.

Collini, C. (1784). Sur quelques zoolithes du Cabinet d'Histoire Naturelle de S.A.S.E. Palatine et de Bavière, à Mannheim. *Acta Academiae Theodoro Palatinae, Mannheim, Pars Physica* **5**: 58-103.

Conard, N. J., Malina, M. (2010). Neue Belege für Malerei aus dem Magdalénien vom Hohle Fels. *Archäologische Ausgrabungen in Baden-Württemberg* **2009**: 52–56.

Conard, N. J., Malina, M. (2011). Neue Eiszeitkunst und weitere Erkenntnisse über das Magdalénien vom

Hohle Fels bei Schelklingen. *Archäologische Ausgrabungen in Baden-Württemberg* **2010**: 56–60.

Crook, A. R. (1894). The lithographic stone quarries of Bavaria, Germany. *Scientific American Supplement* **XXXVIII** (986): 15763-15764.

Crick, R.E. (1983). The practicality of vertical cephalopods shells as paleobathymetric markers. *Geological Society of America Bulletin* **94**: 1109-1116.

Cuvier, G. (1801). Extrait d'un ouvrage sur les espèces de quadrupèdes dont on a trouvé les ossemens dans l'intérieur de la terre. *Journal de Physique, de Chimie et d'Histoire Naturelle* **52**: 253–267.

Cuvier, G. (1809). Mémoire sur le squelette fossile d'un Reptil volant des environs d'Aichstedt, que quelques naturalistes ont pris pour un oiseau, et donc nous formons un genre de Sauriens, sous le nom de Ptero-Dactyle. *Annales du Musée d'Histoire Naturelle, Paris* **13**: 424-437.

Darwin, C. (1859). *On the origin of species by means of natural selection, or the preservation of favoured races in the struggle for life* (1st ed.). London: John Murray.

Döderlein, L. (1923). Anurognathus ammoni ein neuer Flugsaurier. *Sitzungberichte der Bayerische Akademie der Wissenschaften, Mathematisch-Wissenschaftlichen* **1923**: 117-164.

Dzik, J. (1984). Phylogeny of the Nautiloidea. *Palaeontologica Polonica* **45**:1-219.

Ebert, M., Kölbl-Ebert, M., Lane, J.A. (2015). Fauna and Predator-Prey Relationships of Ettling, an Actinopterygian Fish-Dominated Konservat Lagerstätte from the Late Jurassic of Southern Germany. *PLoS ONE* **10**(1): e0116140. doi:10.1371/journal.pone.0116140.

Elgin, R.A., Hone, D.W.E., Frey, E. (2011). The extent of the pterosaur flight membrane. *Acta Palaeontologica Polonica* **56**(1): 99–111.

Elżanowski, A. (2001). A new genus and species for the largest specimen of Archaeopteryx. *Acta Palaeontologica Polonica* **46**(4):519-532.

Elżanowski, A. (2002). Archaeopterygidae (Upper Jurassic of Germany). In L.M. Chiappe., L.M. Witmer (eds). *Mesozoic birds: above the heads of dinosaurs*. Berkley: University of California Press, pp. 129-159.

Enay, R., Bernier, P., Barale, G., Bourseau, J-P., Buffetaut, É., Gaillared, C., Gall, J-C., Wemz, S. (1994). The Ammonites from the Cerin Lithographic Limestones (Ain, France): Stratigraphy and taphonomy. *Geobios* **27**(1): 25-36.

Engust, H. (1961). *Über die Isotopenhäufigkeit des Sauerstoffs und die Meerestemperatur im süddeutchen Malm-delta*. PhD Dissertation: University of Frankfurt.

Evans, S.E., Raia, P., Barbera, C. (2004). New lizards and rhynchocephalians from the Lower Cretaceous of southern Italy. *Acta*

Palaeontologica Polonica **49**(3): 393–408.

Fastnacht, M. (2005). *Jaw mechanics of the pterosaur skull construction and the evolution of toothlessness.* PhD Dissertation: Johannes Gutenberg-Universität in Mainz.

Fastovsky, D. E., Weishampel, D. B. (2009). *Dinosaurs: A concise natural history.* Cambridge: Cambridge University Press.

Faux, C.M., Padian, K. (2007). The opisthotonic posture of vertebrate skeletons: post-mortem contractions or death throes? *Paleobiology* **33**: 201-226.

Feduccia, A. (1993). Evidence from claw geometry indicating arboreal habits of *Archaeopteryx*. *Science* **259**: 790-793.

Feduccia, A. (2012). *Riddle of the Feathered Dragons: Hidden Birds of China.* New Haven, Connecticut: Yale University Press.

Feduccia, A., Tordoff, H. B. (1979). Feathers of Archaeopteryx: Asymmetric vanes indicate aerodynamic function. *Science* **203**(4384): 1021–1022.

Flügel, E. (2013). *Microfacies of Carbonate Rocks: Analysis, Interpretation and Application.* London: Springer.

Foth, C., Haug C., Haug J.T., Tischlinger H., Rauhut O.W.M. (2020) Two of a feather: a comparison of the preserved integument in the juvenile theropod dinosaurs *Sciurumimus* and *Juravenator* from the Kimmeridgian Torleite Formation of southern Germany. In: Foth C., Rauhut O. (eds) *The evolution of feathers: from their origin to the present.* Springer: Cham, pp: 79-101.

Foth, C., Hedrick, B.P., Ezcurra, M.D. (2016) Cranial ontogenetic variation in early saurischians and the role of heterochrony in the diversification of predatory dinosaurs. *PeerJ* **4**:e1589.

Foth, C., Rauhut, O.W.M. (2017). Re-evaluation of the Haarlem Archaeopteryx and the radiation of maniraptoran theropod dinosaurs. *BMC. Evolutionary Biology* **17**(1):236. doi:10.1186/s12862-017-1076-y.

Frey, E., Martill, D. M. (1998). Soft Tissue preservation in a specimen of *Pterodactylus kochi* (Wagner) from the Upper Jurassic of Germany. *Neues Jahrbuch für Geologie und Paläontologie - Abhandlungen* **210**: 421–441.

Frey, E., Tischlinger, H., Buchy, M. C., Martill, D.M. (2003). New specimens of Pterosauria (Reptilia) with soft parts with implications for pterosaurian anatomy and locomotion. *Geological Society, London, Special Publications* **217**: 233–266.

Frey E, Tischlinger H (2012) The Late Jurassic pterosaur *Rhamphorhynchus*, a frequent victim of the ganoid fish *Aspidorhynchus*? *PLoS ONE* **7**(3): e31945.

Fuchs, D. (2007). Coleoid cephalopods from the plattenkalks of the Upper Jurassic of Southern Germany and from the Upper Cretaceous of Lebanon – A faunal comparison. *Neues Jahrbuch Für Geologie Und Paläontologie - Abhandlungen* **245**(1): 59–69.

Fürsich, F.T., Werner, W., Schneider, S., Mäuser, M. (2007). Sedimentology, taphonomy, and palaeoecology of a laminated plattenkalk from the Kimmeridgian of the northern Franconian Alb (southern Germany) *Palaeogeography, Palaeoclimatology, Palaeoecology* **243**: 92-117.

Gaffney, E. S. (1975). A taxonomic revision of the Jurassic turtles Portlandemys and Plesiochelys. *American Museum Novitates* **2574**: 1–19.

Gaillard, C., Bernier, P., Gall, J. C., Gruet, Y., Barale, G., Bourseau, J. P., Buffetaut, E., Wenz, S. 1994. Ichnofabric from the Upper Jurassic lithographic limestone of Cerin, southeast France. *Palaeontology* **37** (2): 285–304.

Gaillard, C., Goy, J., Bernier, P., Bourseau, J.P., Gall, J.C., Barale, G., Buffetaut, E., Wenz, S. (2006). New jellyfish taxa from the Upper Jurassic lithographic limestones if Cerin (France): taphonomy and ecology. *Palaeontology* **49**(6): 1287-1302.

Garofalo, G., Ceriola, L., Gristina, M., Fiorentino, F., Pace, R. (2010). Nurseries, spawning grounds and recruitment of *Octopus vulgaris* in the Strait of Sicily, central Mediterranean Sea. *ICES Journal of Marine Science* **67**: 1363-1371.

Gervais, P. (1871). Remarques au sujet des Reptiles provenant des calcaires lithographiques de Cerin, dans le Bugey, qui sont conservés au Musée de Lyon. *Comptes Rendus des Séances de l'Academie de Sciences* **1871**:79–83.

Gilinsky, N. L., Bennington, J. B. (1994). Estimating numbers of whole individuals from collections of body parts: a taphonomic limitation of the paleontological record. *Paleobiology* **20**(2): 245-258.

Grawe-Baumeister, V.J., Schweigert, G. & Dietl, G. (2000). Echiniden aus dem Nusplinger Plattenkalk (Ober-Kimmeridgium, Südwestdeutschland). *Stuttgarter Beiträge zur Naturkunde Serie B (Geologie und Paläontologie)* **286**: 1-39.

Griffiths, P. J. (1996). The isolated *Archaeopteryx* feather. *Archaeopteryx*. **14**: 1–26.

Groiss, J.R. (1967). Mikropaläeontologische Untersuchungen der Solnhofener Schichten im Gebiet um Eichstätt (Südliche Frankenalb). *Erlanger geologische Abhandlung* **66**: 75-96.

Hallam, A. (2001). A review of the broad pattern of Jurassic sea-level changes and their possible causes in the light of current knowledge. *Palaeogeography, Palaeoclimatology, Palaeoecology* **167**: 23-37.

Hendrickx, C., Hartman, S. A., Mateus, O. (2015). An overview of non-avian theropod discoveries and classification. *PalArch's Journal of Vertebrate Palaeontology* **12**(1): 1-73.

Hoffmann, R., Bestwick, J., Berndt, G., Berndt, R., Fuchs, D., Klug, C. (2020). Pterosaurs ate soft-bodied cephalopods (Coleoidea). *Scientific Reports* **10**(1230): https://doi.org/10.1038/s41598-020-57731-2.

Hoffmann, R., Stevens, K. (2020). The palaeobiology of belemnites – foundation for the interpretation of rostrum geochemistry. *Biological Reviews* **95**: 94-123.

Hone, D. W. E., Henderson, D. M. (2014). The posture of floating pterosaurs: ecological implications for inhabiting marine and freshwater habitats. *Palaeogeography, Palaeoclimatology, Palaeoecology* **394**: 89–98.

Hone, D.W.E., Tischlinger, H., Xu, X., Zhang, F.C. (2010). The extent of the preserved feathers on the four-winged dinosaur *Microraptor gui* under ultraviolet light. *PloS ONE* **5**(2): e9223.

Hoyle, F., Wickramasinghe, N.C., Watkins, R.S. (1985). *Archaeopteryx. British Journal of Photography* **132**: 693-4.

Humphries, S., Bonser, R.H.C., Witton, M.P., Martill, D.M. (2007). Did pterosaurs feed by skimming? Physical modelling and anatomical evaluation of an usual feeding method. *PLoS Biology* **5**: e204.

Huxley, T. H. (1868a). Remarks upon *Archaeopteryx lithographica*. *Proceeding of the Royal Society of London* **16**: 243–48.

Huxley, T.H. (1868b). On the animals which are most nearly intermediate between birds and reptiles. *Annals and Magazine of Natural History* **2**: 66–75.

Huxley, T. H. (1870). Further evidence of the affinity between the dinosaurian reptiles and birds. *Quarterly Journal of the Geological Society* **26**: 32–50.

Jenkyns, H.C., Clayton, C. (1986). Black shales and carbon isotopes in pelagic sediments from the Tethyan Lower Jurassic. *Sedimentology* **33**(1): 87-106.

Ji, Q., Ji, S. (1996). On the discovery of the earliest bird fossil in China and the origin of birds. *Chinese Geology* **233**: 30–33.

Ji, S., Ji, Q., Lu, J., Yuan, C. (2007). A new giant compsognathid dinosaur with long filamentous integuments from Lower Cretaceous of Northeastern China. *Acta Geologica Sinica* **81**(1): 8–15.

Joyce, W. G. (2000). The first complete skeleton of *Solnhofia parsonsi* (Cryptodira, Eurysternidae) from the Upper Jurassic of Germany and its taxonomic implications. *Journal of Paleontology* **74**(4): 684–700.

Joyce, W. G. (2003). A new Late Jurassic turtle specimen and the taxonomy of *Palaeomedusa testa* and *Eurysternum wagleri*. *PaleoBios* **23**(3): 1–8

Joyce, W.G. (2017). A review of the fossil record of basal Mesozoic turtles. *Bulletin of the Peabody Museum of Natural History* **58**(1):65-113.

Jouve, S. (2004). Description of the Skull of a *Ctenochasma* (Pterosauria) from the Latest Jurassic of Eastern France, with a Taxonomic Revision of European Tithonian Pterodactyloidea. *Journal of Vertebrate Paleontology* **24**(3): 542-554.

Kaye, T.G., Pittman, M., Mayr, G.,

Schwarz, D., Xu, X. (2019). Detection of lost calamus challenges identity of isolated *Archaeopteryx* feather. *Scientific Reports* **9**(1182): doi.org/10.1038/s41598-018-37343-7.

Kemp, R.A., Trueman, C.N. (2003) Rare earth elements in Solnhofen biogenic apatite: geochemical clues to the palaeoenvironment. *Sedimentary Geology* **155**: 109–127.

Kemp, R.A., Unwin, D.M. (1997). The skeletal taphonomy of *Archaeopteryx*: a quantitative approach. *Lethaia* **30**(3): 229-238.

Keupp, H. (1977). Der Solnhofener Plattenkalk - ein Blaugrünalgen Lamininit. *Paläontologische Zeitschrift* **51**(1-2):102-116.

Keupp, H., Koch, R., Schweiger, T G., Viohl, G. (2007). Geological history of the Southern Franconian Alb—the area of the Solnhofen Lithographic Limestone. *Neues Jahrbuch für Geologie und Paläontologie, Abhandlungen* **245**: 3–21.

Kidwell, S. M., Baumiller, T. (1990). Experimental disintegration of regular echinoids: roles of temperature, oxygen, and decay thresholds. *Paleobiology* **16**(3): 247-271.

Kieslinger, A. (1939). Revision der Solnhofener Medusen. *Paläeontologische Zeitschrift* **21**: 287-296.

Kin, A., Błażejowski, B. (2012). Polskie Solnhofen. *Przegląd Geologiczny* **60**(7): 375-379.

Kin, A., Błażejowski, B., Binkowski, M. (2012). The 'Polish Solnhofen': a long-awaited alternative? *Geology Today* **28**(3): 91-94.

Kin, A., Gruszczyński, M., Martill, D., Marshall, J.D. & Błazejowski, B. (2013). Palaeoenvironment and taphonomy of a Late Jurassic (Late Tithonian) Lagerstätte from central Poland. *Lethaia* **46**: 71–81.

Kogan, I., Licht, N. (2013). A *Belonostomus tenuirostris* (Actinopterygii: Aspidorhynchidae) from the Late Jurassic of Kelheim (southern Germany) preserved with its last meal. *Paläontologische Zeitschrift* **87**(4):543-548.

Kruta, I., Landmann, N., Rouget, I., Cecca, F., Tafforeau, P. (2011). The role of ammonites in the Mesozoic marine food web revealed by jaw preservation. *Science* **331**(6013): 70-72.

Kuhn, O. (1938). Eine neue Meduse (Hydromeduse) aus dem Oberjura von Solnhofen. *Zoologischer Anzeiger* **122**: 307–312.

Kuhn, O. (1961). Die Tier- und Pflanzenwelt des Solnhofener Schiefers. *Geologica Bavarica* **48**: 1-68.

Lehmann, U., Kulicki, C. (1990). Double function of aptychi (Ammonoidea) as jaw elements and opercula. *Lethaia* **23**: 325-331.

Leinfelder, R.R. (1993). Upper Jurassic reef types and controlling factors: a preliminary report. *Profil* **5**: 1-45.

Longrich, N. (2006). Structure and function of hindlimb feathers in *Archaeopteryx lithographica*. *Paleobiology* **32** (3): 417–431.

Lomax, D. R., Falkingham, P. L., Schweigert, G., Jiménez, A. P. (2017). An 8.5 m long ammonite drag mark from the Upper Jurassic Solnhofen Lithographic Limestones, Germany. *PLoS ONE* **12**(5): e0175426.

Lomax, D. R., Racay, C. A. (2012). A long mortichnial trackway of *Mesolimulus walchi* from the Upper Jurassic Solnhofen Lithographic Limestone near Wintershof, Germany. *Ichnos* **19**: 189–197.

Martill, D.M. (1985). The preservation of marine vertebrates in the Lower Oxfordian Clay (Jurassic) of central England. *Philosophical Transactions of the Royal Society of London. Series B, Biological Sciences* **311**(1148):155-165.

Matyszkiewicz, J. (1996). The significance of *Saccocoma*-calciturbidites for the analysis of the Polish epicontinental late Jurassic Basin: An example from the Southern Cracow-Wielun Upland (Poland). *Facies* **34**: 23-40.

Martyniuk, M.P. (2014). *Beasts of antiquity: stem-birds in the Solnhofen Limestone.* New Jersey: Pan Aves.

Mayr, G. (2017). *Avian evolution: the fossil record of birds and its paleobiological significance.* Chichester: Wiley Blackwell.

Mazzetta, G.V., Christiansen, P., Fariña, R.A. (2004). Giants and bizarres: Body size of some southern South American Cretaceous dinosaurs. *Historical Biology* **16** (2–4): 71–83.

Miethe, A., Born, A. (1928). Die Fluorographie von Fossilien. *Paläontologische Zeitschrift* **9**: 343-356.

Miller, J.E., Pawson, D.L. (1989) *Hansenothuria benti*, new genus, new species (Echinormata: Holothuroidea) from the tropical western Atlantic: a bathyal epibenthic holothurian with swimming abilities. *Proceedings of the Biological Society of Washington* **102**: 977 – 986

Nilsom, C.V. (1994). *Saccocoma*: a benthic crinoid from the Jurassic Solnhofen Limestone, Germany. *Palaeontology* **37**(1): 121-129

Müller, M.K. (2011). The fish fauna of the Late Jurassic Solothurn Turtle Limestone (NW Switzerland). *Swiss Journal of Geosciences* **104**(1): 133-146.

Newman, E. (1843). Note on the Pterodactyle Tribe considered as marsupial bats. *Zoologist* **1**:129-131.

Nichols, G.J. (2009). *Sedimentology and Stratigraphy* (2nd Ed.) London: Wiley-Blackwell.

Nudds, R. L., Dyke, G. J. (2010). Narrow primary feather rachises in *Confuciusornis* and *Archaeopteryx* suggest poor flight ability. *Science* **328** (5980): 887–889.

O'Connor, P., Claessens, L. (2005). Basic avian pulmonary design and flow-through ventilation in non-avian theropod dinosaurs. *Nature* **436** (7048): 253–256.

Olsen, P.E., Kent, D.V., Sues, H.D., Koeberl, C., Huber, H., Montanari, A., Rainforth, E.C., Fowell, S.J., Szajna, M.J., Hartline, B.W. (2002). Ascent of dinosaurs linked to an iridium anomaly at the Triassic-Jurassic boundary. *Science* **292**(5571):1305-1307.

Olson, S. L.; Feduccia, A. (1979). Flight capability and the pectoral girdle
of *Archaeopteryx*. *Nature* **278** (5701): 247–248.

Ősi, A. (2010). Feeding-related characters in basal pterosaurs: implications for jaw mechanisms, dental function and diet. *Lethaia* **44**: 136-152.

Ostrom, J.H. (1970). *Archaeopteryx*: notice of a "new" specimen. *Science* **170**(3957):537-538.

Ostrom, J. (1974). *Archaeopteryx* and the origin of flight. *The Quarterly Review of Biology* **49**(1):27-47.

Ostrom, J. H. (1969). Osteology of *Deinonychus antirrhopus*, an unusual theropod from the Lower Cretaceous of Montana. *Bulletin of the Peabody Museum of Natural History* **30**: 1-165.

Ostrom, J. H. (1973). The ancestry of birds. *Nature* **242**(5393): 136.

Ostrom, J. H. (1976). *Archaeopteryx* and the origin of birds. *Biological Journal of the Linnean Society* **8** (2): 91-182.

Pan, Y., Fürsich, F.T., Chellouche, P., Hu, L. (2019). Taphonomy of fish concentrations from the Upper Jurassic Solnhofen Plattenkalk of Southern Germany. *Neues Jahrbuch für Geologie und Paläontologie - Abhandlungen* **292**(1): 73-92.

Pawson, D.L. (1982). Deep sea echinoderms in the tongue of the ocean, Bahamas Islands: a survey, using the research submersible Alvin. *Australian Museum Memoir* **16**: 129-145

Perry, L. (2014). Fossil Focus: Annelids. *Palaeontology[online]* **4**(11):1-8.

Petit, G., Khalloufi, B. (2012). Paleopathology of a fossil fish from the Solnhofen Lagerstätte (Upper Jurassic, southern Germany). *International Journal of Paleopathology* **2**: 42-44.

Peyer, Karin (2006). A reconsideration of *Compsognathus* from the upper Tithonian of Canjuers, southeastern France. *Journal of Vertebrate Paleontology* **26**(4): 879–896.

Pike, A. V. L., Maitland, D. P. (2006). Scaling of bird claws. *Journal of Zoology* **262**(1): 73-81.

Rauhut, O. W. M., Foth, C., Tischlinger, H., Norell, M. A. (2012). Exceptionally preserved juvenile megalosauroid theropod dinosaur with filamentous integument from the Late Jurassic of Germany. *Proceedings of the National Academy of Sciences* **109** (29): 11746–11751.

Reisdorf, A.G., Wuttke, M. (2012). Re-evaluating Moodie's opisthotonic-postG. (ure hypothesis in fossil vertebrates' part I: reptiles—the taphonomy of the bipedal dinosaurs *Compsognathus longipes* and *Juravenator starki* from the Solnhofen Archipelago (Jurassic, Germany). *Palaeobiodiversity and Palaeoenvironments* **92**(1): 119-168.

Reolid, M., Mattioli, E., Nieto-Albert, L.M., Rodríguez-Tovar, F.J. (2014). The Early Toarcian Oceanic

Anoxic Event in the External Subbetic (Southiberian Palaeomargin, Westernmost Tethys): Geochemistry, nannofossils and ichnology. *Palaeogeography Palaeoclimatology Palaeoecology* **411**:79–94.

Ruben, J.A., Jones, T.D., Geist, N.R., Hillenius, W.J. (1997). Lung structure and ventilation in theropod dinosaurs and early birds. *Science* **278**(5341): 1267–1270.

Rudkin, D. M., Young, G. A., Nowlan, G. S. (2008). The oldest horseshoe crab: a new xiphosurid from Late Ordovician Konservat-Lagerstätten deposits, Manitoba, Canada. *Palaeontology* **51**(1): 1-9.

Rudloe, A. (1980). The breeding behavior and patterns of movement of horseshoe crabs, Limulus polyphemus, in the vicinity of breeding beaches in Apalachee Bay, Florida. *Estuaries* **3**(3):177-183.

Schäfer, W. (1972). *Ecology and palaeoecology of marine environments*. Chicago: The University of Chicago Press.

Schmitz, L., Motani, R. (2011). Nocturnality in dinosaurs inferred from scleral ring and orbit morphology. *Science* **332**(6030): 705-708.

Schwark, L., Vliex, M., Schaeffer, P. (1998). Geochemical characterisation of Malm Zeta laminated carbonates from the Franconian Alb, SW-Germany (II). *Organic Geochemistry* **29**: 1921– 1952.

Schwarz-Wings, D., Klein, N., Neumann, C., Resch, U. (2011). A new partial skeleton of Alligatorellus (Crocodyliformes) associated with echinoids from the Late Jurassic (Tithonian) lithographic limestone of Kelheim, S-Germany. *Fossil Record* **14** (2):195–205.

Schweigert, G. (1998a). Die Spurenfauna des Nusplinger Plattenkalks (Oberjura, Schwäbische Alb). *Stuttgarter Beiträge zur Naturkunde, Serie B* **262**: 1–47.

Schweigert, G. (1998b). Ein Stück "Festland" im Solnhofener Plattenkalk. *Jahresberichte und Mitteilungen des Oberrheinischen Geologischen Vereins* **80**: 271–278.

Schweigert, G. (2007). Ammonite biostratigraphy as a tool for dating Upper Jurassic lithographic limestone from South Germany –first results and open questions. *Neues Jahrbuch für Geologie und Paläeontologie* **245**(1): 117-125.

Schweigert, G. (2009). First three-dimensionally preserved in situ record of an aptychophoran ammonite jaw apparatus in the Jurassic and discussion of the function of aptychi. *Berliner paläobiologische Abhandlungen* **10**: 321-330.

Schweigert, G., Maxwell, E., Dietl, G. (2016). First record of a true Mortichnium produced by a fish. *Ichnos* **23**(1-2): 71–76.

Schweigert, G., Salamon, M., Dietl, G. (2008). *Millericrinus milleri* (SCHLOTHEIM, 1823) (Crinoidea: Millericrinida) from the Nusplingen Lithographic Limestone (Upper Kimmeridgian, SW Germany). *Neues Jahrbuch für Geologie und Paläontologie - Abhandlungen.* **247**: 1-7.

Seilacher A. (1982). Ammonite shells as habitats — floats or benthic islands? (Abstract). In: Einsele G., Seilacher A. (eds) *Cyclic and Event Stratification*. Berlin: Springer.

Seilacher, A. (1993). Ammonite aptychi: How to transform a jaw into an opercilum. *American Journal of Science* **293**(A): 20-32.

Seilacher, A. (2007). *Trace fossil analysis*. Springer: Berlin.

Seilacher, A., Reif, W., Westphal, F. (1985). Sedimentological, ecological and temporal patterns of fossil Lagerstätten. *Philosophical Transactions of the Royal Society B* **311**: 5-23.

Selden, P., Nudds, J. (2012). *Evolution of fossil ecosystems* (2nd ed). London: Manson Publishing.

Senter, P. (2006). Scapular orientation in theropods and basal birds and the origin of flapping flight. *Acta Palaeontologica Polonica* **51**(2): 305–313.

Senter, P., Robins, J.H. (2003). Taxonomic status of the specimens of *Archaeopteryx*. *Journal of Vertebrate Paleontology* **23**(4): 961-965.

Sereno, P.C., Martinez, R.N., Wilson, J.A., Varricchio, D.J., Alcober, O.A. (2008). Evidence for avian intrathoracic air sacs in a new predatory dinosaur from Argentina. *PLoS ONE* **3**(9): e3303doi=10.1371/journal.pone.0003303.

Shipman, P. (1981). *Life history of a fossil: an introduction to taphonomy and paleoecology*. Cambridge, MA: Harvard University Press.

Simões, T.R., Caldwell, M.W., Nydam, R.L., Jiménez-Huidobro, P. (2017). Osteology, phylogeny, and functional morphology of two Jurassic lizard species and the early evolution of scansoriality in geckoes. *Zoological Journal of the Linnean Society* **180**(1): 216–241.

Simpson, G.C. (1926). Are *Dromatherium* and *Microconodon* mammals? *Science* **63**: 548-549.

Speakman, J. R., Thomson, S. C. (1994). Flight capabilities of *Archaeopteryx*. *Nature* **370** (6490): 514.

Stevens, K., Mutterlose, J., Schweigert, G. (2014). Belemnite ecology and the environment of the Nusplingen Plattenkalk (Late Jurassic, southern Germany): evidence from stable isotope data. *Lethaia* **47**: 512-523.

Stewart, F.J., Cavanaugh, C.M. (2006). Bacterial endosymbioses in *Solemya* (Mollusca: Bivalvia)--model systems for studies of symbiont-host adaptation. *Antonie Van Leeuwenhoek* **90**(4): 343-360.

Stow, D. (2010). *Vanished ocean: how Tethys shaped the world*. Oxford: Oxford University Press.

Sutton, M., Perales-Raya, C., Gilbert, I. (2015). A phylogeny of fossil and living neocoleoid cephalopods. *Cladistics* **32**(3): 297-307

Swinburne, N.H.M., Hemleben, C. (1994). The plattenkalk facies: a deposit of several environments. *Geobios* **27**(1): 313-320.

Tanabe, K., Inazumi, A., Tamahama, K., Katsuta, T. (1984). Taphonomy of half and compressed ammonites from the Lower Jurassic black shales of the Toyora area, west Japan. *Palaeogeography, Palaeoclimatology, Palaeoecology* **47**(3-4): 329-346.

Tintori, A. (1992). Fish taphonomy and Triassic anoxic basins from the Alps: a case history. *Rivista Italiana di paleontogia e stratigrafia* **97**(3-4): 393-408.

Tischlinger, H. (1998). Erstnachweis von Pigmentfarben bei Plattenkalk-Teleosteern. *Archaeopteryx* **16**: 1–18.

Tischlinger, H. (2010). Pterosaurs of the "Solnhofen" Limestone: new discoveries and the impact of changing quarrying practises. IActs Geoscientica Sinica *31*(1): 76-78

Tischlinger, H., Arratia, G. (2013). Ultraviolet light as a tool for investigating Mesozoic fishes, with a focus on the ichthyofauna of the Solnhofen archipelago. In Arratia, G., Schultze, H.P., Wilson, M.V.H (eds.). *Mesozoic Fishes 5 – Global Diversity and Evolution.* Munich: Verlag Dr. Friedrich Pfeil.

Unwin, D.M. (2005). *Pterosaurs from deep time.* New York: Pi Press.

Urey, H., Lownestam, H.A., Epstein, S., McKinney. (1951). Measurement of palaeotemperatures and temperatures of the Upper Cretaceous of England, Denmark, and the Southeastern United States. *Bulletin of the Geological Society of America* **62**: 399-416.

van Iten, H., Marques, A.C., Leme, J.d.M., Pacheco, MLAF, Simões, M.G. (2014). Origin and early diversification of the phylum Cnidaria Verrill: major developments in the analysis of the taxon's Proterozoic–Cambrian history. *Palaeontology* **57**: 677-690.

van Straaten, L.M.J.U. (1978). Dendrites. *Journal of the Geological Society* **135**: 137-151.

Vidovic, S.U., Martill, D.M. (2014). *Pterodactylus scolopaciceps* Meyer, 1860 (Pterosauria, Pterodactyloidea) from the Upper Jurassic of Bavaria, Germany: The Problem of Cryptic Pterosaur Taxa in Early Ontogeny. *PLoS One* **10**(2): e0118015.

Viohl, G. (1996). The Paleoenvironment of the Late Jurassic fishes from the southern Franconian Alb (Bavaria, Germany). In Arratia, G., Viohl, G (eds.). *Mesozoic Fishes 1- Systematics and Paleoecology.* Munich: Verlag Dr. Friedrich Pfeil.

Voeten, DFAE, Cubo, J., de Margerie, E., Röper, M., Beyrand, V., Bureš, S., Tafforeau, P., Sanchez, S. (2018). Wing bone geometry reveals active flight in Archaeopteryx". *Nature Communications* **9** (923): doi:10.1038/s41467-018-03296-8.

von Huene, F. (1926). Vollständige Osteologie eines Plateosauriden aus dem schwäbischen Keuper. *Geologische und Paläontologische Abhandlungen, Neue Folge* **15** (2): 139–179.

von Meyer, H. (1837). Mitteilung an Prof. Bronn (*Plateosaurus engelhardti*). *Neues Jahrbuch für Geologie und Paläontologie* **1837**: 316.

von Meyer, H. (1839). *Idiochelys fitzingeri*, eine Schildkröte aus dem Kalkschiefer von Kelheim. *Beiträge Zur Petrefacten-Kunde* **1**: 59–74.

von Meyer, H. (1861). *Archaeopterix lithographica* (Vogel-Feder) und Pterodactylus von Solenhofen. *Neues Jahrbuch für Mineralogie, Geognosie, Geologie und Petrefakten-Kunde* **1861**: 678–679.

von Meyer, H. (1862). *Archaeopteryx lithographica* aus dem lithographischen Schiefer von Solenhofen. *Palaeontographica* 10: 53–56.

von Meyer, H. (1864). *Parachelys eichstättensis* aus dem lithographischen Schiefer von Eichstätt. *Palaeontographica* **11**: 289–295.

von Zittel, K. A., Eastman, C. R. (1913). *Text Book of Paleontology*. London: MacMillan & Co

Wagler, J. (1830). *Natürliches System der Amphibien* Munich, 1830: 1–354

Wagner, J. A. (1859). Über einige im lithographischen Schiefer neu aufgefundene Schildkröten und Saurier. *Gelehrte Anzeigen der Bayerischen Akademie der Wissenschaften* **49**: 553.

Wagner, J. A. (1861). Neue Beiträge zur Kenntnis der urweltlichen Fauna des lithographischen Schiefers; V. *Compsognathus longipes* Wagner. *Abhandlungen der Bayerischen Akademie der Wissenschaften* **9**: 30–38.

Wellnhofer, P. (1967). Ein Schildkrötenrest (Thalassemydidae) aus den Solnhofener Plattenkalken. *Mittelungen der Bayerischen Staatssamlung für Paläontologie und Historische Geologie* **7**: 181–192.

Wellnhofer, P. (1970). Die Pterodactyloidea (Pterosauria) der Oberjura-Plattenkalke Suddeutschlands. *Bayerische Akademie der Wissenschaften, Mathematisch-Wissenschaftlichen Klasse, Abhandlungen* **141**: 1-133.

Wellnhofer, P. (1975). Die Rhamphorhynchoidea (Pterosauria) der Oberjura-Plattenkalke Süddeutschlands. II. Systematische Beschreibung. *Palaeontographica A* **148**: 132–186.

Wellnhofer, P. (1978): *Handbuch der Paläoherpetologie - Teil 19: Pterosauria*. Stuttgart: Gustav Fischer Verlag.

Wellnhofer, P. (1988). A New Specimen of *Archaeopteryx*. *Science* **240** (4860): 1790–1790.

Wellnhofer, P. (1991). *The Illustrated Encyclopedia of Pterosaurs*. New York: Crescent Books

Wellnhofer, P. (2009a). A short history of pterosaur research. *Zitteliana* **29**: 7-19.

Wellnhofer, P. (2009b). *Archaeopteryx: the icon of evolution*. Munich: Verlag Dr. Friedrich Pfeil.

Wenz, S., Bernier, P., Barale, G., Bourseau, J.P., Buffetaut, E., Gaillard, C., Gall, J-G. (1994). The ichthyofauna from the lithographic limestones from the lower Kimmeridgian of Cerin (Var, France). *Geobios* **27**. (sup. 1): 61–70.

Wilby, P. R., Brigg S, D. E. G., Viohl, G. (1995). Controls on the phosphatization of soft tissues in

Plattenkalks. *International Symposium on Lithographic Limestones, 9–16th July, 1995, Lleida-Cuenca, Spain. Extended abstracts, II. Ediciones Universidad Autonóma de Madrid*:165–166.

Wilkin, J.T.R. (2018). The geological history and paleoenvironment of the south German Plattenkalks. *Fossil News* **21**(1):44-51.

Wilkin, J.T.R. (2019a). Taphonomy of Tithonian fishes from the Mörnsheim Formation of southern Germany. *Zitteliana* **93**:81-85.

Wilkin, J.T.R. (2019b). Use of ultraviolet light in Plattenkalk research. *SSP Blog (European Geoscience Union)*. Available: https://blogs.egu.eu/divisions/ssp/2019/12/16/use-of-ultraviolet-light-in-plattenkalk-research/

Wilkin, J.T.R. (2020). The south German Plattenkalks. *Geology Today* **36**(1):28-33.

Wilson, M.V.H. (1989). Taphonomic processes: information loss and information gain. *Geoscience Canada* **15**(2): 131-148.

Witton, M.P. (2013). *Pterosaurs: natural history, evolution, anatomy*. Princeton: Princeton University Press.

Witton, M. P. (2018). Pterosaurs in Mesozoic food webs: a review of fossil evidence. In: Hone, D. W. E., Witton, M. P. & Martill, D. M (eds). *New Perspectives on Pterosaur Palaeobiology*. London: Geological Society Special Publications **455**: 7-23.

Wolf, S., Dapschauskas, R., Velliky, E., Floss, H., Kandel, A. W., Condrd, N. J. (2018). The use of ochre and painting during the Upper Paleolithic of the Swabian Jura in the context of the development of ochre use in Africa and Europe. *Open Archaeology* **4**: 185-205.

Xu, X.; Norell, M.A. (2004). A new troodontid dinosaur from China with avian-like sleeping posture. *Nature* **431** (7010): 838–841.

Zachos, J.C., Dickens, G.R., Zeebe, R.E. (2008). An early Cenozoic perspective on greenhouse warming and carbon-cycle dynamics. *Nature* **45** (7176): 279-283.

Zell, P., Beckmann, S., Stinnesbeck, W. (2014). *Liostrea roemeri* (Ostreida, Bivalvia) attached to Upper Jurassic ammonites of northeastern Mexico. *Palaeobiodiversity and Palaeoenvironments* **94**: 439–451.

CPSIA information can be obtained
at www.ICGtesting.com
Printed in the USA
BVHW062126180521
607553BV00012B/1966